Rules, Reason, and Self-Knowledge

Rules, Reason, and Self-Knowledge

Julia Tanney

HARVARD UNIVERSITY PRESS
Cambridge, Massachusetts · London, England
2013

Printed in the United States of America

Library of Congress Cataloging-in-Publication Data

Tanney, Julia, 1959–
Rules, reason, and self-knowledge / Julia Tanney.
p. cm.
Includes bibliographical references and index.
ISBN 978-0-674-06708-0 (alk. paper)
1. Philosophy of mind. 2. Cognitive science. I. Title.
BD418.3.T36 2013
128'.2—dc23 2012012113

For Crispin

Contents

Rules, Reason, and Self-Knowledge

Introduction

THE ESSAYS collected in this volume, written over the last two decades, paint a picture of rules, reason, and self-knowledge—in short, the mind—that emerges from a philosophical puzzle whose solution requires taking a close look at the initial assumptions that are responsible for its generation. Unpicking an assumption in one area tends to lead to knots in others, and the tangles they generate need to be unravelled in turn.

This unsnarling amounts to a sustained criticism of today's canon in philosophy of mind: a collection of views about mental phenomena and explanation that give rise to the mind-body problem and its now widely accepted "solution" in functionalism and its offshoots in representationalism and computationalism. The essays in this volume chronicle a journey away from a position that I no longer occupy. The way I think about the mind has changed as some of the basic assumptions, often embedded in the technical terms of contemporary philosophy of mind, are questioned and rejected as I travel. The philosophical *imprimatur* for these assumptions is to be found, among others, in the work of the early Davidson, Putnam, and Fodor. The view I come to turns out to be the one they were so keen to rebut—one that has clear affinities with the work of Wittgenstein and Ryle.

This is the view that the normative dimension along which our various doings, sayings, thinkings, and perceivings can be assessed has to be understood, not as normative properties of mental states—the nature and explanatory role of which remains mysterious—but as embedded in descriptions that are themselves sophistications upon ever ascending levels of discourse, or sociolinguistic practices. Taken seriously sixty years ago, this position is today invisible to those doing cutting-edge work at the intersection of philosophy of mind, language, and moral psychology, especially as these subjects join hands with empirical research that is conducted along lines championed in the cognitive sciences.

The topics in these essays crisscross and overlap, but I have imposed a structure on them under four key themes. Within these sections, the essays appear in the order in which they were written. Since each was conceived to be free standing, there is unavoidable repetition; with a few exceptions, the alterations I have made are stylistic or for enhanced clarity.

I. Rules and Normativity

The journey begins with philosophical puzzles of irrationality. When trying to spell out what is irrational about self-deception and weakness of the will, we find ourselves articulating ideals of thought and action: "One ought not to believe what one has overwhelming evidence against"; and "once all the options have been considered, the one that 'wins through' in deliberation should be implemented in action." These are examples of norms or principles of rationality. How are we to understand their nature and their role?

Let us take as our starting point the now traditional story in the philosophy of mind which, inherited from Descartes, comes to us in its modern incarnation via (among others) Davidson and Fodor. According to this view, reasons are content-bearing mental states which both rationalize and cause the action they explain. The motivation for "individualizing" reasons in this way arises from the intuition that an agent may have a reason for acting in a particular way but not accept it as her own, and (the thought goes) unless she does, it cannot provide an explanation of her action. This whole approach will be challenged, especially in chapters 5–9, but let us accept it for the moment. Given this understanding, how are we to construe the status of the norms that govern reasoning?

Davidson suggests that just as a person cannot be said to have acted for a reason unless it is the one *for which* she acts, a person cannot be accused of violating a norm unless she, in some sense, accepts it. An individual's acceptance of the norm, for Davidson, amounts to its figuring as a content of thought: e.g., "I ought to act in accordance with my all things considered judgment." This, however, creates a problem for him. For if the norm figures as a judgment-content, it should, by his lights, determine or be causally sufficient for the action which it prescribes. But if the norm, so construed, determines action that is in accord, how can rendering it as the content of judgment help us understand norm violation?

Something has clearly gone wrong, and in chapter 1, "De-Individualizing Norms of Rationality," I argue that Davidson's proposed solution does not work, not because it needs another patch, but because the problem is misconceived at the outset. Persons—rational agents—cannot simply escape the requirements of rationality by failing to accept its norms. We cannot make sense of our interpretive-psychological practices—those in which we attribute reasons and ascribe propositional-attitudes—unless those norms are *somehow* in place. At the end of the chapter I suggest that there is some sort of level, type, or category error in attributing norms of rationality to the subject as the contents of her (causally efficacious, content-bearing) mental states.

The theme that persons cannot simply opt out of the norms that govern thought and action is revisited in chapter 2, "Normativity and Thought," my Joint Session response to David Papineau. Papineau is right, I argue, to recognize that his brand of naturalism is threatened by the idea that norms govern thought and judgment, for he wishes to account for mental phenomena in nonsemantic and nonnormative terms. This reductionist project, though, fails to allow for the normative dimension along which our doings and deliberatings can be assessed. Papineau attempts to block the threat of normativity by sequestering the norms to an area where they do no damage: to the contents of the agent's desires. According to his view, one is only subject to norms insofar as one desires to be. Papineau's move, like Davidson's, attempts to individualize the norms of rationality by construing them as ingredients in an individual's psychology. I have another attempt at showing this cannot work. We cannot make sense of belief-desire psychology if the remit of the norms is conditioned by the individual's acquiescence. We cannot even make sense of this acquiescence, nor of the agent's beliefs, desires, or other propositional attitudes unless these norms of rationality are in some sense presupposed.

Presupposed, but how? Indeed, what about the obvious fact that they are violable? Irrationality was, after all, our starting point—accepting the genuineness of this phenomenon is what prompted us to articulate norms of thought and action in the first place. This suggests that the norms do not (merely) figure as ingredients in one's psychology as contents of mental states but rather as part of the "background" in some way.

In chapter 3, "Playing the Rule-Following Game," I consider generalizations of the problem. It is not only the norms of rationality that present us with a puzzle. Logical norms, as Lewis Carroll had already noted, also lead to paradox if it is assumed that they figure as ingredients within the logical practices they in some sense govern. To construe the norms by which we assess a person's doings, thinkings, and sayings as encoded in rules which are supposed to guide her is the first step in what we may call the "individualist's dilemma," which leads to two pernicious horns. The idea behind individualism is that in order to be credited with thinking logically, acting rationally, speaking meaningfully, and so forth, an individual must not only act in accordance with the norms by which we assess her; she must also be aware of the norms and in a strong sense guided by them. She would be guided by them in this sense if the norms were encoded in rule-representations which she consults in the course of acting. When the norms under investigation are broadly construable as rational norms, this temptation leads to a regress because the activity of following a rule presupposes the ability to act rationally. Call this the "intellectualist" or "cognitivist" horn of the dilemma. Individualists who recognize the threat of regress hope to avert it by construing the grasp and implementation of the rules as embodied either in subpersonal mental states or in subpersonal causal mechanisms. Insofar as subpersonal states are supposed to be "content-bearing," this maneuver inherits the regress difficulties in supposing the rules are followed normally; the other maneuver, with its focus on mechanisms, fails to deliver the sort of distance necessary between the norms and the moves (in reasoning, language, thinking, perception) they govern so that the former can be used as standards by which to assess the latter. Call this the "norm-obliterating" horn.

We will see later how this dilemma affects the view that reasons are content-bearing, causally efficacious states. (Indeed, it even threatens the view that beliefs, desires, and other propositional attitudes have vehicles: *a fortiori,* ones in the head.) For now, let us look at the status of the norms and the rule-representations in which they are embodied. Language, thought, and action are each implicated in the activity of following a rule. So they are abilities that must be presupposed, and not explained by rule-following, at least if we are to understand this activity as one of consultation and obeyance. It seems that this argument generalizes to all the abilities that are presupposed in rule-following: reading the representation of the rule, understanding what it requires, and implementing it in action. Norms of thought and action may indeed figure in our thoughts: I, in writing this, and you, in following me, are presumably thinking about them now. When they figure in thoughts, however, they cannot in the same sense figure in an account of the constitutive nature of meaning, thought, or rationality. They must already be "in the picture."

Chapter 4, "Real Rules," spells out how this is so, and I speak up here on Wittgenstein's behalf against what I consider to be a misinterpretation by Crispin Wright of his remarks on rule-following. Norms are abstracted from sociolinguistic practices that are up and running; codifications of these norms may be articulated in rules by a theorist (for the purposes of teaching, training, and evaluating) in order to provide the numerous dimensions along which her doings, sayings, thinkings, and perceptions (under various descriptions of ascending "thickness") may be assessed. Looking at the role of norms in this light, I argue, makes metaphysical quandaries about rules and their applications disappear.

The individualist's dilemma derives from the conviction that in order to be credited with thinking logically, acting rationally, speaking meaningfully, and so on, an individual must not only act in accordance with the norms that govern these practices; she must also be aware of those norms and guided by them. The latter requirement, as I see it, is the key idea behind cognitivism. To avoid the dilemma we must relax our requirements for membership to these rational, linguistic practices.

II. Reason-Explanation and Mental Causation

Davidson's individualism, we saw, was motivated by the thought that an individual may have a reason for a particular action but not accept it as her own. In such a case, even though it provides an "anaemic" justification, the reason cannot provide an explanation of the action. It is this general form of argument that prompts Davidson to suppose that reasons must do more than rationalize action if they are to explain it: reasons must cause it as well. Reasons are, on this view, construed as ingredients in an individual's psychology. Davidson's arguments to this effect, and for the thesis that reason-explanation is a species of causal explanation, were (in spite of his own misgivings) considered by most philosophers to clear the way for an empirical, science-inspired study of the mind.

An untenable metaphysics of the mind, however, replete with insurmountable puzzles, is unleashed by Davidson's argument. If one accepts with him and others that causation is an empirically discoverable, contingent relation between events that have (whether or not we know them) logically independent descriptions, then a particular understanding of the nature of reasons will be difficult to resist. This is the view that reasons are, or consist in, a combination of mental states or events, whose ontological nature is to be investigated, but which must (given certain plausible assumptions) have some kind of dependence upon, or realization in, the agent's brain. Individualism, given certain compelling commitments, quickly metamorphoses into internalism.

The chapters in this volume, especially 5–9, present a sustained attack on this picture. The conclusion is the conditional one that if we were to accept these requirements on causation, the explanatory power of propositional attitude ascriptions or of reason-explanation is not to be found in causal relations between mental "states," thus construed. Mental causation exists, of course—as do mental states. But the explanatory role of propositional attitude ascriptions, I argue, functions other than to name states or triggering-events that figure in the production of behavior described under the relevant action-type.

Davidson's individualism construes reasons as belief-desire pairs; his internalism takes beliefs, desires, and other cognitive and pro-attitude states to be event-tokens in the brain. Some of the difficulties with this view, which are crystallized in the problems of mental causation, the explanatory role of the mental, and the threat of epiphenomenalism, are examined in chapter 5, "Why Reasons May Not Be Causes." On the one hand, causation is deemed by Davidson to be necessary for reason-explanation. On the other, his views on causation, causal explanation, and the anomalousness of the mental suggest that a causal *explanation* may never be achieved, for we will probably never know the laws that subsume the *relata* (whose physical descriptions may also elude us) because laws govern events only as physically described. These views are in tension. If it is possible or even likely that we will never know the laws, or the physical description of the events they subsume, they can hardly be necessary for reason-explanation. This undermines the motivation, I argue, for insisting that causation figures in an account of it in the first place.

What are reasons, then, if not some combination of vehicle and content: the former somehow realized in complex causal interactions in the brain, the latter whose realizations may also reach out to the physical world beyond? When one wonders how to act, one may reflect on reasons; and certainly, our practice of determining reasons is inextricably linked with that of attributing beliefs and desires. In chapter 6, "Reason-Explanation and the Contents of the Mind," I explore the relation between reason-explanation and propositional attitude ascription. A point that is often missed is that reason-explanation may be secured by adverting to information that has nothing to do with the agent's "attitudes." A woman's fleeing the building is explained by its being on fire. "Did she believe it to be on fire?" needs special stage-setting in order to reveal its appositeness or its purpose as a question. If the circumstances were such that she may not have known the building was on fire, her beliefs would be relevant. If, in that situation, her flight from the building was mysterious given her belief that it was on fire, her particular desires or more details about the circumstances would be pertinent as well. Although Davidson is right that desires and beliefs have some important, conceptual relation to action, rather than taking as our starting point the agent's "mental states," which are supposed to play a causal role that "mirrors" their semantic one, I suggest that we begin with reason-explanation. The acknowledgment that an individual has beliefs and desires that cohere with her action, as described, is to confirm that we have pinpointed an overall sense-making pattern—a dimension along which her doings can be assessed. Doing this, however, does not require that we construe her beliefs and desires as independent existents that play a role in the production of behavior.

This is one attempt to rework the "logical connection" argument: in such a way, incidentally, that there should be no temptation to associate this view with philosophical or analytic behaviorism. In chapter 7, "Reasons as Non-Causal, Context-Placing Explanations," I return to the source of this argument by considering its articulation by Melden, with some assistance from Anscombe and Ryle. Davidson's argument for the causal-explanatory role of reasons famously alludes to Melden's driver who raises his hand to signal a turn. The fact that he raises his hand may, in those circumstances, constitute his signaling, Davidson agrees, but nonetheless he may have raised his hand for another reason: say, to wave to a friend.

We can concede this. But this does not show, as Davidson argues, that a causal relation is required between reason and action in order to secure explanation. Reprising an idea originally put forward in "Why Reasons May Not Be Causes," I argue that it merely shows that what was one candidate for a sense-making pattern should be rejected for another. Davidson and Fodor seem to construe the logical connection argument as setting conditions on the appropriate descriptions that are used in reason-explanation. That is, they seem to construe Melden as insisting that causal *relata* have to be described so that they evince no logical connection. This absurd interpretation of the view, which

Melden warned against, would prohibit us from saying the sun caused the sunburn. The correct reading of the logical connection argument is, I maintain, that for a causal relation to obtain (of the kind that tempts us to postulate (the triggering) of mental states or vehicles as *relata*) the events must be independently describable in principle. But the existence of such independently describable events—the alleged referents of the propositional attitude ascription—is precisely what is in question.

To attribute propositional attitudes, I suggest, is not to name a state of the agent the nature of which threatens to be mysterious unless we can construe it as emerging from her brain and the brain's causal relations to the environment. It is rather to put a marker down on a particular pattern of descriptions of her thoughts, actions, and sayings which puts them into an understandable context. The ability to see actions as fitting into familiar patterns comes about through training and shared forms of life: a kind of acculturation which enables us to see actions in new ways. There is no way of articulating this "sense-making pattern" or "context which renders her doings, sayings, thinkings, understandable," other than by issuing reminders about what we would say in the particular circumstances illustrated. Indeed, one of the central themes of these essays is that the ability to see these patterns, contexts, or circumstances as sense-making ones comes about through acculturation or training but is not acquired through prior theoretical, rule-following operations; the ability to follow a rule presupposes the ability to recognize such patterns.

Chapter 8, "Pain, Polio, and Pride: Some Reflections on 'Becausal' Explanations" defends this way of looking at the role of mental concepts against the objections of (early) Putnam who was partly responsible for dislodging the view that mental explanation of action is, in a sense that is ill-understood, conceptual. Putnam suggested that we understand mental concepts such as pain by analogy with what he took to be natural kind concepts such as Polio or Multiple Sclerosis. Mental concepts "fix the reference" of mental states, whose nature it is for the psychologist (or neurologist) to discover. I argue that this analogy is misconceived for Putnam's purposes, though useful for mine. For Poliomyelitis was, like diseases such as AIDS and CFS, a syndrome; these diseases form a subset of non-causal, context-placing explanations. This does not, of course, prevent us from identifying one or more viral or bacterial origins, which in turn may enable us to uncover an underlying microstructural explanation for the diseases' effects. But classifying something as a syndrome or disease does not serve as a mere placeholder for that of a microstructural explanation: it plays a sense-making or explanatory role of its own. Diseases, understood as syndromes, have something therefore in common with explanations invoking character traits such as pride. In this chapter, I begin to explore and question some of the philosophical assumptions about language, description, and explanation that lurk in the background of Putnam's argument.

III. Philosophical Elucidation and Cognitive Science

The model of the mind that is my target here—one I think fundamentally mistaken—has infiltrated not only subbranches of philosophy but figures as the nuts and bolts of the cognitivist trend and its attendant "scientism" that has had such a powerful impact on the humanities and social sciences.

How did cognitivism come to have such appeal? Are its philosophical foundations really stable? In chapter 9, "How to Resist Mental Representations" (a critical review of Tim Crane's *The Mechanical Mind*), I trace the path that has led philosophers to posit mental representations, as well as their reasons for embracing the view that causally efficacious, content-bearing mental states are the referents of our ordinary propositional attitude terms. In attempting to resist the central claims of Crane's book, which lead to a defense of computationalism, I show how "the problem of error" and the "normativity of the mental" are part of the same phenomenon, and that to "mechanize" explanation, as the causalists and cognitivists wish to do, obliterates the explanatory framework that is essential to reason-explanation and the commonsense psychological practices in which it figures; it is to become impaled on the "norm-obliterating" horn of the individualist's dilemma. This should resonate with those who have been impressed by the arguments of Kripke's Wittgenstein: we agree (at least) that the norms embodied in meaning rules (indeed, in any phenomena constitutive of rationality) cannot be located within the causal-dispositional mechanisms of an individual's psychology.

The study of mental phenomena, I claim, involves looking at the way expressions containing mental concepts are correctly used and how, in these various uses, they function. Choices and decisions may of course be made about what should count as correct use; this will change according to our purposes for such classification. There is nothing in this story that prevents our making decisions to adopt one use or reject another: just as disease concepts may change in the light of discoveries of microstructural origins or cognate symptoms, the cartography of our mental concepts is fluid. Mental concepts have "jobs" which can be discerned from investigating the familiar practices in which they figure. Thus, any suggestion to change their contours would need to be considered in the light of the work they are already doing.

This idea is taken up in the context of a popular philosophical thought experiment in chapter 10, "On the Conceptual, Psychological, and Moral Status of Zombies, Swamp-Beings, and Other 'Behaviorally Indistinguishable' Creatures." There is something troubling, indeed, morally disturbing in the idea that consciousness concepts (and therefore, some think, all mental concepts) can, let alone should, be withheld from creatures who are behaviorally indistinguishable from us in all possible circumstances, yet who lack something ineffable, which can only be gestured at in the philosophical jargon terms of "conscious experience" or "qualia." The same can be said, I argue, about our spontaneously conceived doppelgängers whose supposed inner representational states have not evolved. (The normative nature of the mental is thus, I imply, not to be found in the teleosemanticist's notion of *proper function*.) It is not simply, as Dennett argues, that the debate is silly on the grounds that the metaphysical possibility of such creatures is indefinitely small. It is rather that such a view would completely change the meaning of our consciousness concepts. Since concepts are not, I believe, fixed in advance of the linguistic practices in which they can be discerned as abstractible features, and since they are fluid not only within these practices, but as these practices change, I demur at describing zombies and swamp men as conceptually impossible. Those who suggest, however, that there is no internal inconsistency in the possibility of such creatures are making a radical break with how the terms "expressing" consciousness concepts are used. The burden—including not only an appeal to explanation but also to morality—is on them to show why we should accept their proposed revision. This discussion is not an attempt to resurrect behaviorism, which I (along, incidentally, with Wittgenstein and Ryle) reject. Various considerations regarding the circumstances might incline us to withhold mental descriptions that in other circumstances we would bestow, given the same "behavior" (in some "thin" sense of the word). My aim is rather to remind the reader of the conceptual dependence (of a kind I attempt to tease out in several chapters of this volume) between the mind and its expressions in actions, speech, and behavior that is inherent in our use of mental concepts.

What of the standard riposte that some mental phenomena—mental representations for example—are theoretical posits that owe their allegiance to the scientific practices in which they are embedded and not to the "intuitions of a small group of idiosyncratic people"? This question is addressed in chapter 11, "Conceptual Analysis, Theory Construction, and Philosophical Elucidation in the Philosophy of Mind." The term "mental representation" moonlights between different jobs: on the one hand it seems to be used as an alternative for "propositional attitude," where this is considered to name a content-bearing state; on the other, it is thought to belong to a widely held but nonetheless controversial theory about the nature of these states. Little matter, I argue. Both uses of "mental representation" germinate from propositional attitude *ascriptions* and are thus constrained by the jobs these ascriptions perform. I make clear my doubt that such ascriptions perform their roles by naming, individuating, or referring to states with content. (Incidentally, this alone puts me at odds with contemporary theorists, even those—diverging in part from the Davidson-Fodorian ancestry which I take as my target—who take it as given that mental predicates function to name items—events or states—the (ontological) nature of which is a matter of debate.) The false supposition that they do refer to such states infects second-order accounts or theories about their nature, so—inappropriate talk of philosophers' "intuitions" aside—the admittedly technical use of "mental representations" is not quarantined against considered judgments about what constitutes the correct application of propositional attitude predicates.

Chapter 12, "Ryle's Regress and the Philosophy of Cognitive Science," returns to Fodor's early arguments, which, along with Davidson's, were largely responsible for reintroducing the Cartesian framework—individualism, internalism, and mental causal explanation—that is the target of this volume. Ryle's regress objection to the "Intellectualist Legend"—that intelligent activity requires prior theoretical operations—was recognized by Fodor to present a powerful conceptual obstacle to the premise that underlies cognitivist approaches in the social sciences. Fodor attempts to thwart Ryle's argument in *The Language of Thought* by accusing him of confusing causal and conceptual explanations and claiming that, by analogy with computers, we can see how the appeal to explicit rules ends at the first-order "program" level. Second-order rules are reducible to built-in causal processes, thus the regress is halted. Fodor's arguments against Ryle, however, fail for two reasons. First, Fodor's appeal to the "empirical necessity" of theoretical operations misfires because he is the one who has misunderstood the difference between causal and conceptual questions. Second, reprising discussions in earlier chapters, I argue that the fact that second-order rules are reducible to causal processes shows, not that the regress is halted, but that we cannot consider intelligent activity by analogy with computers.

IV. Self-Knowledge

The picture of the mind that emerges, so far, is this. To understand the nature of the mind and mental phenomena requires understanding mental concepts. To investigate concepts, technical or nontechnical, is to study the employments of expressions—sentences in use—from which concepts can be discerned as abstractible features. To investigate *belief, desire, intention, action,* and *reason* is to study the jobs—the explanatory work—of expressions containing the relevant terms in the practices in which they are deployed. These mental concepts begin life as nontechnical terms; suggestions to revise them should be considered in the light of how well they do their jobs in order to gauge whether, under the proposed revision, these performances can be improved.

Most contemporary theorizing, however, simply assumes, falsely, that mental terms function as those such as "table" or "brain": to name a manifest (type of) entity the constitution of which is open to philosophical/theoretical speculation. Theories of the mind that take mental concepts to pick out states—even content-bearing ones—that are alleged to emerge, supervene upon, or be realized in individuals' brains make a mistake *about language* which is not importantly different from Descartes'. Though it may seem antiscience, the view I am commending is best seen as a reaction against a too-crude philosophical theory of meaning. Nonetheless, it is one that dominates the present-day philosophical scene and has been accepted by many of those working in mainstream philosophy as well as in the "mental sciences" today.

In this Introduction, mainly for ease of exposition, I have drawn attention to the sense-making patterns which we, as observers, trace or construct in describing the actions of, and ascribing propositional attitudes and other mental concepts to, the one we are trying to fathom. This way of putting it, however, masks a complexity that attends the interpretation of a language-speaker, for the sense-making pattern crucial for understanding such a subject will include her sayings as a subset of her doings. She vents her feelings, proclaims her intentions, announces and defends her beliefs, professes her desires, weighs aloud her reasons, and above all, describes her own actions in talking about herself.

The norms of rationality associated with thought and action are presupposed by our interpretive practices; the self-ascriber is not immune to them. This idea resonates throughout this volume, for a ubiquitous theme is that when the goal of interpretation involves understanding the doings and sayings of a self-reflective agent, she, too, is subject to the sense-making standards that govern our psychological-ascriptive practices.

This is particularly clear when a subject's avowals and self-ascriptions puzzle us because they seem at odds with our own understanding. In an early essay (slightly revised for this volume) I suggest that there is something to be learned from looking at the (albeit rare) cases in which someone *does* have to think about what she thinks or how she feels and perhaps reflect upon her dreams, fantasies, flashes of imagery, memories, surges of emotion, and sensations. My point of departure in chapter 13, (retitled) "Some Constructivist Thoughts About Self-Knowledge," is Crispin Wright's suggestion that "the authority which our self-ascriptions . . . assume is not based on any kind of cognitive advantage, expertise, or achievement"; "it is . . . the subject's right to declare what he intends . . . and his possession of this right consists in the conferral upon such declarations, other things being equal, of a *constitutive* rather than *descriptive* role." It seems to me that caution is needed in understanding what is right about this intriguing suggestion, for whatever element of creativity is enjoyed by the one who self-pronounces, her use of mental concepts in these activities is subject to the same constraints as her use of them in other, third-person ascriptions. I tease out the sense in which both "description" and "invention" or "construction" are apposite expressions to mark this particular (sense-making) aspect of self-expression and ascription.

This idea is explored further in chapter 14, "Self-Knowledge, Normativity, and Construction." The traditional Cartesian view, with its commitment to the essential privacy of the whole range of mental phenomena, not only of sensations, after-images, and dreams, but also of thoughts, beliefs, and desires, fails to accommodate the standards to which we are accountable when we express, avow, or self-ascribe, some of which ascriptions, of course, reach beyond what can be "seen" by a ghostly spectator of a mental theatre. Philosophies of mind that inherit aspects of the Cartesian picture—those that construe the range of mental phenomena as (at bottom) inner, causally effective states—will, I suspect, have difficultly accommodating the "constructivist" element of ascription: the feature of the agent's self-conception or "practical identity" that plays a partly constitutive role in determining what she believes, thinks, means, and values. Rather than focusing on the functionalist literature on self-knowledge, I consider instead Richard Moran's more nuanced arguments against the view that the relation between self-conception and emotions or other first-order mental states is causal rather than logical, individuative, or constitutive. Stressing, again, the sense-making norms to which self-ascribers are subject I develop the thought that, to borrow an expression David Wiggins uses in discussing ethical discourse (but imported into a context of which he would, I believe, not approve), the "compossibility of objectivity, discovery, and invention" is *also* part of our ordinary (nontheoretical and nonscientific) understanding of the mind.

In order, provisionally, to understand the picture that is emerging, let us expand the "sense-making" canvas to include two salient patterns that coincide in some places, but not in others. We put an agent's actions, including her sayings, into understandable contexts that are marked out by a pattern of intentions, beliefs, meanings, and thoughts that makes sense of her doings. She, as a self-reflecting agent, does the same in attributing reasons, thoughts, and meanings to herself. Sometimes, perhaps most of the time, what she says in trying to make sense of herself reveals what we may agree is an acceptable sense-making path. At times, however, it does not. When it does not, we may reject it for another that we prefer. The pattern she has articulated, however, nonetheless reveals something important about her: namely, her self-conception. Her self-conception, whether or not it coincides with what we think, will tend in turn to influence what she does.

This picture, though it may be helpful, is nonetheless a simplification: expressions that embody self-conceptions are not limited to reflective self-interpretations in adult, language-using human beings. For what are we to make of avowals, which are not, by definition, instances of self-reflection? They do not result from puzzling over how one should make sense of oneself; as Ryle pointed out, to avow "I feel depressed" is not a piece of scientific premise-providing, but a piece of conversational moping. The natural thing to do is to "speak one's mind" and to refrain from doing so is learned, sophisticated behavior. Speaking one's mind is best understood, he suggests, not as a mode of "discovery" but rather as voluntary nonconcealment.

The phenomenon of expressing one's feelings, thoughts, or intentions is puzzling precisely because on the one hand, in expressing herself the avower seems to be doing something which is not, in general, making a statement which can be challenged on the grounds of its truth or falsity, but rather by virtue of its truthfulness, genuineness, or sincerity. On the other hand, not only does the avowal look grammatically like a statement, but the first-person employment of the same sentence in a different (interpretive) context may function as such—indeed, as a statement which is precisely challengeable on the grounds of its truth or falsity. Add to this the thought that my concurrence or disagreement with your avowal, "yes, you are depressed" seems, on the face of it, to treat the correlative noninterpretive outburst as truth-evaluable after all. The first consideration speaks in favor of a view of the mind in which the act of avowing is not descriptive or reportive: the second and third, of a view in which it is.

It seems to me that Expressivists are right to recognize what Ryle calls the preposterousness of the assumption that every true or false statement either asserts or denies that a mentioned object or set of objects possesses a specific attribute. Nonetheless, they undermine this insight by presuming exactly the opposite: that mental predicates function *in general* to attribute to an individual a state of mind—now construed by them as "expressed" in the act of avowal. This move is a disaster because the obvious question, "what is the nature of this state?" becomes both pressing and impossible to answer. In chapter 15, "Speaking One's Mind" (an expanded version of my critical review of Dorit Bar-On's *Speaking My Mind*), I argue that Bar-On's "account" of avowals fails to make headway in the traditional problem of self-knowledge because she accepts the picture of "mental states" that leads to the quandary in the first place: the problem, that is, of how an avower can be credited with knowledge of a state when the grounds for accepting a claim about its existence depends upon what she does or goes on to do, sometimes long in the future. That this condition will be fulfilled is not something she is in a position to avow.

The progression along the interpretive scale from vented feelings to self-ascriptions is a fluid one for language-speaking, social creatures. There is no easy demarcation between what a person believes, means, thinks, and intends and what she should believe, mean, think, and intend—at least where the "should" is qualified by "if we are to understand her." This is one sense in which the "normativity" of the mental reaches bedrock: there is no room, when it comes to the domain of the mental, for what some might construe as the "merely" descriptive.

What, then, is it to have thoughts, beliefs, or desires? What is it to act for reasons? These questions cannot be answered independently of how we come to know what a person thinks, believes, or wants, or how we come to know her reasons, and these in turn cannot be answered without investigating how we make decisions about ascribing mental predicates. The sense-making patterns to which I allude, like the concepts we use to mark them, will change their contours depending upon the circumstances and our purposes. There are different routes for ascribing such predicates, which sometimes converge and sometimes do not. Whether someone has a particular belief or intention depends on which particular "inflections of meaning" or "elasticities of significance," to use Ryle's expressions, we (tacitly) accept on the occasion we consider the question.

In chapter 8, "Pain, Polio, and Pride," I made some suggestions as to how we can accommodate this picture for sensation-concepts which, because "episodic," fit squarely within the domain of the avowable and yet still have to meet objective standards for their application. In the final chapter of this volume, "Conceptual Amorphousness, Reasons, and Causes," I revisit the notion of mental causation as well as the relation between avowals and self-ascriptions in our interpretive practices. I illustrate how expressions that are not ambiguous nonetheless have inflections of meaning which affect the implication and evidential threads they withstand and support. Unlike ambiguous words, these have a common root, or family resemblance. But the root or resemblance they share is not one that allows us, independently of the circumstances in which the expression is deployed, to discern its logical contours. This has radical implications for mental concepts and thus for our understanding of the mind.

There are different things we count as the correct application of a mental predicate. On the one hand we look at what the person says about herself; on the other, we look at her nonlinguistic, as well as linguistic actions and behavior. We normally make assumptions about how an agent conceives her actions, and this assumption conditions our understanding, indeed our description, of them. We assume that people intend to do what they do, that they mean what they say, and that what they say about their performances will accord with the sense that we make of them. We are not ordinarily aware of this, since our very description of the performance already contains an assumption that the agent stands in this relation to it. It is only when this assumption is questioned or rejected—and we have to reassess or redescribe the action—that we become aware of the role this assumption plays. When this conviction is shaken, how do we reassess the relation between the agent and her action? It depends: sometimes we are interested in what makes most sense, regardless of what she says; sometimes we are interested in her self-conception. Sometimes, we can make no easy division between the two. Various considerations, circumstances, and purposes affect how we ascribe to others and how we treat their avowals and self-ascriptions.

Inflections of meaning give concepts the power to express an indefinite variety of ideas; it is only when they are studied in the particular circumstances of their use, expressed in sentences performing their particular jobs, that we can understand their logical force. Failing to allow for the complexity of mental discourse, we are tempted to suppose its terms function to pick out or name objects with extraordinarily complex properties: a mental mechanism, which, in its contemporary incarnation is assumed to be realized in the brain. Wittgenstein identified this temptation in his Cambridge lectures in the 1930s, and Ryle presented a systematic critique of it in 1949. The tendency to mechanize the mind should be resisted. Accepting the power of dictions to express an indefinite number of ideas and of sentences to perform an indefinite number of jobs, will, I suggest, not only rid us of metaphysical quandaries; as it happens, it will also enable us to avoid paradox in our quest to understand irrationality.

PART ONE

Rules and Normativity

CHAPTER ONE

De-Individualizing Norms of Rationality

1. It seems to be a platitude that what makes behavior irrational is its failure to accord with some particular norm of rationality, and it seems right to say that intentional action by and large conforms to these norms. These considerations might encourage one to attempt to explain an individual's ability to act rationally, and account for some of her lapses, by attributing to her "knowledge"—either explicit or tacit—of what the norms require. The norms of rationality in some sense govern thought and action. But is the sense in which they do this captured by construing them as psychologically internalized rules, or as causal determinants of behavior?

The need to attribute some particular principle of rationality to an individual is defended by Davidson explicitly in his characterization of akrasia.[1] I should like to explore his attempt to "individualize" the principle, or render it into a norm which is "cognized" by the individual whose actions are governed by it. This will require taking some space

1. A correlative move—the need to internalize reasons—is made in Davidson's causal account of intentional action. In "Why Reasons May Not Be Causes" (chapter 5), I consider how the arguments I give here affect the thesis that reasons are causes.

to explicate Davidson's causal account of intentional action, which, for the sake of making the arguments clear, I shall just accept. I shall show that it is not necessary to individualize a principle of rationality in order to characterize an individual's actions as internally irrational. In the second half of the chapter I shall develop this argument by considering in detail what explanatory role an individual's cognitive grasp of such norms might play. I shall argue that there is no construal of "cognitive grasp" such that attributing cognitivist grasp of a norm to an individual would explain her dispositions to act in accordance with what the norm prescribes, either directly, or via her second-order explicational abilities. I argue in the end that "cognizing" a norm of rationality could only be considered constitutive of an individual's ability to *obey* it on a very artificial and stipulative sense of "obey." I conclude that it is a mistake to construe the principles of rationality as norms or rules which may or may not be obeyed or followed.[2]

2. In "Actions, Reasons, and Causes," Davidson argues that rationalizing explanations, which explain an agent's actions by citing his reasons for doing what he does, are a species of causal explanations.[3] We can specify the mental cause of the action—the reason for which an agent acts as he does—by citing a pro-attitude or desire the agent

---

2. I have attempted to draw out in more detail how Davidson's strategy is subject to the problems suggested by Baier and maintained her line against defenses suggested by Pears. A detailed exploration of these suggestions is important, since many of the ideas defended by Davidson—notably, the idea that reasons are causes—(which, agreeing with Baier, I think is the real culprit in rendering irrationality a paradox) have been dominant (again) in the philosophy of mind since the mid 1960s, largely because of Davidson's arguments. The intuition that there is something odd about this thesis, and perhaps something correspondingly odd about the research programs that depend on it, needs to be developed in such a way that those who do not already share the intuition might be persuaded of its truth. See Annette Baier, "Rhyme and Reason: Reflections on Davidson's Version of Having Reasons," and David Pears, "Rhyme and Reason—Response to Baier," *Actions and Events—Perspectives on the Philosophy of Donald Davidson,* Ernest Lepore and Brian McLaughlin, eds. (Oxford: Blackwell, 1985), 116–129 and 130–135.

3. Donald Davidson, "Actions, Reasons, and Causes," (1963), *Essays on Actions and Events* (Oxford: Clarendon Press, 1980), 3–20.

has towards actions of this kind (or toward the events or states brought about by actions of this kind), and a belief that the action is of this kind or will bring about the desired outcome.

Davidson suggests that the logical relationship between the contents of reasons and actions can be described as a practical syllogism: the pro-attitude and belief that cause the action (my desire for adventure, and my belief that spending the weekend in Barcelona will enable me to satisfy this desire) have contents that can be described as premises in an argument the conclusion of which is a description of an intention that corresponds to, or prescribes, an action. Thus the desire-constituent of a reason (say, my reason for spending the weekend in Barcelona) can be expressed as: "Any act of mine which is likely to yield adventure is desirable." When coupled with the belief premise "spending the weekend in Barcelona is likely to yield adventure," it follows that any act of mine which is my spending the weekend in Barcelona is one I may judge to be desirable. The problem with this view is that any act-type which is likely to yield adventure will be one whose desirability follows from my pro-attitude. In "How is Weakness of the Will Possible?" Davidson proposes that the propositional content of the desire-component should be changed to express what he calls a "prima facie" evaluation (or judgment) which qualifies event- or action-types: any action of mine is desirable insofar as it is likely to yield adventure.[4] This, together with a premise expressing the belief component "spending the weekend in Barcelona is likely to yield adventure," implies a judgment which must itself be relativized as the major premise is above: "spending the weekend in Barcelona is desirable insofar as it is likely to yield adventure." But this kind of judgment is too weak to determine action: although spending the weekend in Barcelona is desirable insofar as it is likely to yield adventure, it might be undesirable for other reasons. In this article Davidson claims that the action itself must correspond to something stronger than a prima

4. Donald Davidson, "How is Weakness of the Will Possible?," (1969), *Essays on Actions and Events*, 21–42.

facie evaluation that the act is desirable in a certain respect; it must correspond to an unconditional or "all-out," singular judgment expressing the desirability of a particular action.[5]

The upshot of this view is that there will be a logical gap, according to Davidson, between the content of prima facie evaluations and the content of all-out judgments. The contents of my reasons for acting deductively imply the content of a relativized judgment on my part, that insofar as the act would enable me to satisfy a particular desire, performing it is desirable. But they do not deductively imply an all-out, "de-relativized" judgment that I ought to perform the action (say, that I ought to spend the weekend in Barcelona). And yet, this sort of all-out judgment just is the intention to act that Davidson holds to be a necessary concomitant to action.

Prima facie evaluations that conflict, like desires that conflict, must compete in order to be realized in action, so the psychological model needs to be extended to capture the logical structure between judgments made in deliberation: the relationship between judgments that adjudicate between conflicting prima facie evaluations. This structure is evinced in deliberation insofar as choices are made about which reasons are to subserve others. If these choices are rational then according to standard theories of decision, and arguably common sense, deliberation proceeds by one's ranking or weighting certain desires or judgments higher than others on the basis of one's (rational) grasp of their relative importance and the probability of the possible outcomes.

Although deliberation (the mental process) perhaps only occurs self-consciously, a self-conscious weighting of alternatives need not be required for intentional action, or for actions to be rationalized. All that is required is that the explanatory relations between reasons, judgments about the relative ranking of reasons, and intentions that seem to be exhibited in self-conscious deliberation be attributable to the subject

5. That an all-out judgment about the desirability of the action is necessarily connected to an act performed intentionally is implicit here; the connection between all-out judgment and intending is made in Donald Davidson, "Intending," (1978), *Essays on Actions and Events*, 83–102. For criticism, see David Pears, *Motivated Irrationality* (Oxford: Clarendon Press, 1984).

as part of our rationalizing project. Thus we may say that, although not self-consciously deliberating, the agent has nonetheless acted intentionally. And this seems undeniable. There is manifest rationality in all sorts of "automatic" human behavior which is not a product of self-conscious deliberation.[6]

In "How is Weakness of the Will Possible?" Davidson suggests that the judgment that manifests the relative ranking of reasons—i.e., the result of deliberation—is an "all things considered judgment." An all things considered judgment is "doubly relativized." First it is relativized according to the way in which the desire would be satisfied in the commission of the action (say, as in the prima facie judgment: "spending the weekend in Barcelona is desirable insofar as it is likely to yield adventure"). It is relativized also according to its place with respect to other desires and in the light of the agent's beliefs, principles, and values. This judgment might be something like: "In the light of my ranking the opportunity for adventure over prudential concerns, and in the light of my beliefs about what spending the weekend in Barcelona will involve, and so on, spending the weekend in Barcelona is desirable." According to Davidson, this all things considered judgment is conditional in form and thus, like the singly relativized judgments that logically precede it, does not entail the kind of judgment which is a necessary concomitant to intentional action. Again, this latter judgment—which Davidson identifies as an intention—must be unconditional, or de-relativized. So the logical gap that exists between the contents of prima facie evaluations, or sentences describing them, and the contents of intentions, or sentences describing them, is still preserved on the extended model between all things considered judgments and actions. The move from

---

6. This move plays a crucial role in the argument and in similar arguments of the kind. It is to insist on the explanatory propriety of attributing intentional properties in a way which may not correspond to the subject's (self-) explicational abilities as manifested either in her avowals or in her acts of self-interpretation. Burge makes an analogous move in his anti-individualism arguments when he insists on the explanatory propriety of attributing concepts to an individual whose concept-explicational abilities are incomplete. Note that Davidson sanctions this move in "Reply to Bratman," *Essays on Davidson: Actions and Events,* Bruce Vermazen and M. Hintikka, eds. (Oxford: Clarendon Press, 1985), 195–201.

a doubly relativized judgment like "Assuming that I have considered all relevant things, I ought to spend the weekend in Barcelona" to an unconditional (de-relativized) judgment like "I ought to spend the weekend in Barcelona" is not a move that is prescribed by first-order logic since, presumably, some piece of relevant information not considered might always defeat the claim that I ought to spend the weekend in Barcelona. Thus, the failure to make such a move in one's thinking cannot (yet) be taken to exhibit a kind of logical inconsistency.

It is precisely this kind of failure that Davidson takes to be exhibited in paradigmatic akratic action: "the action of an agent who, having weighed up the reasons on both sides, and having judged that the preponderance of reasons is on one side, then acts against this judgment." Since this type of irrationality is, for Davidson, *ineliminable,* the introduction of the all things considered judgment is a propitious addition to the account of intentional action since it allows the logical space for akrasia. But leaving the logical space for irrationality makes characterizing the error that is manifested in it that much more difficult. This is apparent to Davidson who wonders, "since logic cannot tell me which to do, it is unclear in what respect either action would be irrational."[7]

This sets up the major program of "Paradoxes of Irrationality," which is to diagnose which principle has been violated in akrasia, and—the thesis I shall be challenging—to characterize the norm violation intra-individually or intra-psychically so that it is reflected as an inconsistency within (descriptions of) the agent's mental states or events (including relevant descriptions of action). It is not obvious straight off how this strategy is to succeed: if practical reasoning is not straightforwardly explicable in terms of logical coherence among contentful states, then

7. Donald Davidson, "Paradoxes of Irrationality," (1982), *Problems of Rationality* (Oxford: Clarendon Press, 2004), 169–188. This is the other crucial move of the argument and arguments of its kind: it is to insist on the explanatory propriety of attributing intentional properties in a way which manifests a kind of internal error. In this case actions are identified as akratic or irrational, in such a way that they are nevertheless tokens of intentional action. Burge makes an analogous move in his anti-individualism argument when he argues for the propriety of attributing concepts to individuals whose application of them is partially mistaken.

it is not obvious that surd practical reasoning can be straightforwardly explicated in terms of logical incoherence among contentful states. But worse: if such a general account were to be given, then since the right kind of logical relations are, according to Davidson, in some sense constitutive of the very thoughts and actions we are using in our descriptions of irrational states or events, there is a constant worry that the kind of inconsistency needed to characterize irrationality will at the same time undercut the identification of the mental events that are used in its description. This, then, is the paradox of irrationality. Anyone attempting to give a general account of irrationality has to show how descriptions of irrational events are possible given the apparent constitutive nature of the principles that are violated.

The burden of the discussion that follows is to explore the way in which these principles are constitutive.

3. Before offering an account of what goes wrong in akratic action, Davidson gives an example of straightforward intentional action, in which a person acts for reasons:

> A man walking in a park stumbles on a branch in the path. Thinking the branch may endanger others, he picks it up and throws it in a hedge beside the path. On his way home it occurs to him that the branch may be projecting from the hedge and so still be a threat to unwary walkers. He gets off the tram he is on, returns to the park, and restores the branch to its original position. Here everything the agent does (except stumble) is done for a reason, a reason in the light of which the corresponding action was reasonable, for example, given that he wants to remove the branch and he believes that getting off the tram will enable him to remove the branch, it was reasonable for him to get off the tram.[8]

Now consider additional information which renders the action akratic. When the man returns to the park to remove the branch, although he

8. Davidson, "Paradoxes of Irrationality," 172–173.

has a reason for getting off the tram, say he has a better reason (not wasting the time) to continue on it. Suppose the man accepts that he has a better reason to stay on the tram, and that he judges, all things considered, that he ought to continue. What needs to be explained, according to Davidson, is not why he gets off the tram (we saw that he has a reason for that) but rather why he does not act according to his better judgment.

Davidson's suggestion as to how to characterize the kind of irrationality that occurs in the paradigmatic case of akrasia is to posit a principle of practical reasoning which will bridge the logical gap between the penultimate outcome of deliberation—namely, an all things considered judgment—and the intention itself. The principle says that I ought to act in accordance with my all things considered judgment (and form de-relativized judgments or intentions consistently with it). Davidson calls this a "second-order" principle, presumably because it speaks about the deliberation process itself, and does not necessarily get bandied around within it. Introducing the second-order principle allows Davidson to diagnose what goes wrong in a case of akrasia by pinpointing the norm that has been flouted.

I would like to take a closer look at Davidson's introduction of the second-order principle. For the sake of argument I shall accept it as diagnostic of one kind of irrationality. I would like to explore what an individual agent's relationship to it is supposed to be, and what work this relationship is doing in the characterization of akrasia. What follows is Davidson's formal description of what he takes to be a paradigmatic case of akratic action.

> Pure internal inconsistency enters only if I also hold—as in fact I do—that I ought to act on my own best judgment, what I judge best or obligatory, everything considered. [. . .] A purely formal description of what is irrational in an akratic act is, then, that the agent goes against his own second-order principle that he ought to act on what he holds to be best, everything considered. It is only when we can describe his action in just this way that there is a puzzle about explaining it. If the agent does not have the principle that he ought to act on what he holds best, everything considered,

> then though his action may be irrational from our point of view, it need not be irrational from his point of view—at least not in a way that poses a problem for explanation. For to explain his behaviour we need only say that his desire to do what he held to be best, all things considered, was not as strong as his desire to do something else.[9]

What does it mean to "hold" the principle? In what sense does the agent "have" the principle that he act on what he holds best, everything considered? When Davidson says that in acting akratically the agent goes against his own second-order principle, he seems to be suggesting that the principle be attributed as the content of one of the subject's own mental states.

Now it might be thought that in insisting that the paradox of irrationality only arises for individuals who "hold" the principle their actions violate, Davidson is attempting to call our attention to a distinction invoked in moral psychology between *external* and *internal* irrationality. A person is externally irrational if she violates the norms of her community in choosing some end, but internally rational if she engages in appropriate means-ends reasoning to achieve it. Thus a person might be externally irrational insofar as she wants to die, but internally rational insofar as her suicide plans suit the intended goal. The idea seems to be that unless the principle of rationality is attributable to an individual as one of her ends, her violating it cannot be construed as problematic from her own point of view. But this reasoning is odd, since attributing the principle as an end presupposes the individual to whom it is

9. "Paradoxes of Irrationality," 177. At times, it seems that Davidson *stipulates* that the "paradigmatic" akratic action that he wants to analyze necessitates attributing the norm violated to the akrates, when for example, he says: "The standard case of akrasia is one in which the agent knows what he is doing, and why, and knows that it is not for the best, and knows why. He acknowledges his own irrationality." At other times as in the quote cited in the text above, it seems he is *arguing* for the necessity of the attribution. In any case, I hope to return a negative verdict to a possibility he contemplates at the end of this article: "I have urged that a certain scheme of analysis applies to important cases of irrationality. Possibly some version of this scheme will be found in every case of "internal" inconsistency or irrationality." "Paradoxes of Irrationality," 186.

attributed is already disposed to act in accordance with it, and if she is, it is not clear what attributing it to her as an end would explain.[10]

And yet, clearly the principles of rationality may be possible objects of thought: I am presumably thinking about them now as I write. But what of the akrates? Perhaps it makes sense to attribute to him knowledge of the principle if he is able to explicate his actions in the light of it: to diagnose his akratic actions as irrational, to correct his actions which violate it, or to justify those which accord with it. And perhaps it might be thought that having these second-order abilities is necessary for him to be internally irrational. But, intuitively, this does not seem to be true.

To bring this point home, and to see how external irrationality with respect to the principle of continence tends to collapse into internal irrationality, suppose the agent we are considering has not got the ability to correct his actions in the light of perceived violations of the norm, he has not got the ability to justify his actions in the light of it, etc. Then the purported explanation of the action of this individual—which Davidson claims would prevent its characterization as (internally) irrational—is that "his desire to do what he held best all things considered, wasn't as strong as his desire to return to the park." But how would appealing to the strengths of the desires circumvent a diagnosis of internal irrationality? On a simple model of deliberation, whatever difference to the outcome of deliberation the strength of a desire makes, it makes before the all things considered judgment is formulated. If the desire to return to the park is stronger, it ought to be the survivor of deliberation and championed by the all things considered judgment. If instead the desire to stay on the tram survives

10. I argue for this in detail in section 4. In later articles Davidson suggests that no one can be interpretable as failing to hold the principle, but he fails to see this as a threat to his argument for requiring that the principle be attributed to the individual in order to diagnose akrasia. If no one can be interpretable as failing to hold the principle, then, as I shall argue in section 5, the norm has metamorphized into a *law,* and there would be no reason anymore to attribute the principle to any particular individual (a fortiori as part of any particular individual's "cognitive equipment"). See "Incoherence and Irrationality" (1985) and "Deception and Division" (1986), *Problems of Rationality* (189–198 and 199–212.

deliberation, then presumably this is because it was stronger. But all this is true of any deliberator: even one who is not able to correct himself when he fails to act in accordance with his all things considered judgment. Thus it is not clear how adverting to the strength of the desire in the case where the principle cannot be attributed as a representation allows us to cancel a diagnosis of irrationality; the fact of a particular desire's strength must already have been taken into account by the time the all things considered judgment was made. In fact, this is the same point that Davidson makes when he describes the error in deliberation for one who does hold the principle of continence ("the desire to return to the park entered twice over"), only here, supposing we can imagine an individual who does not have the explicational abilities, we can pinpoint the same kind of failure; thus whether or not he holds the principle of continence is immaterial. Again: that a desire overpowers an all things considered judgment that issues from a contest that the desire itself was too weak to win evinces some kind of error. That it does manifest an error is what Davidson himself points to in describing the fault that occurs in his paradigmatic akratic act. And yet its being an error does not depend on the agent's being able to correct himself in the light of failing to act as his deliberations dictate, etc. The implication that we can accept *as not (internally) irrational* the action of one who does not have these abilities is wrong.[11]

The upshot is that attributing the principle to an agent as an object of knowledge or as a cognitive state is not necessary to diagnose internal irrationality. Because supposing someone does *not* have the second-order explicational abilities (which attributing knowledge of the principle ostensibly would explain), how could we *ever* escape the conclusion that nonetheless, given his status as an agent, a delibera-

11. A number of people have suggested that on a more complex model of deliberation, this "fault" can be modeled so as to allow the possibility of desires surviving in action, despite deliberation to the contrary. But this misses the point. The question is not whether the phenomenon (in which the "strongest" desires survive in action despite deliberation to the contrary) is *possible;* the question is whether or not the phenomenon is *irrational.* My claim is simply that if it is irrational, we do not need to attribute cognition of the norm that is violated to the subject in order to see it as an internal error.

tor, a practical reasoner, etc., he ought—*by his own lights*—to correct himself in the light of his understanding that he has acted against his better judgment? After all, what is the point of his deliberating if he is not going to act in accordance with his deliberations? Indeed, why would he get as far along in the deliberation process as to reach the all things considered judgment if he will not act in accordance with it? The argument is that on the hypothesis that an individual has not got the second-order abilities to correct and justify her actions in the light of the principle, she is nevertheless acting *internally* irrationally when her actions fail to conform to it.

I would like to backtrack now, and explore the premises in the argument above in more detail. I began this essay by stating two intuitions. First, that it seems to be a platitude that irrationality is the violation of some norm of rationality. And second, that the principles of rationality in some sense govern thought and action.

Perhaps we need the principles as diagnostic tools, but if the above argument about the principle of continence is correct (and generalizable to other principles of rationality) then it would seem unnecessary to attribute them as objects of cognition to the agent we are attempting to diagnose or teach.

The argument rests on the premise that in order to be attributed as objects of cognition, the norms must be serving some explanatory role in the cognitive/psychological life of the individual to whom cognitive grasp of the norm is attributed. I shall argue in detail in what follows that there is no *normative* role for the cognized principle to play.

4. It might be thought that attributing knowledge of a norm to an individual would explain the individual's disposition to act in accordance with it. After all, as Davidson supposes, the akrates must be disposed to act in accordance with this norm, even though on occasion he violates it.

But the fact that one acts in conformity with a norm or principle does not yet give a reason to attribute knowledge of it to the agent. Your resting in your chair instead of floating away manifests behavior that accords with the principles of physics, but attributing to you knowledge of these principles does not explain why you do not float away. Of course, it might explain other things, for example, how you are able to answer a question correctly on a physics exam about why you do not float away.

Although this is a point which arises over and over again it will be worth making the problem explicit in its present incarnation.[12] Crispin Wright describes the intuitive idea of how norms or rules might figure in a psychological explanation:

> Correctly applying a rule to a new case will, it is natural to think, typically involve a double success: it is necessary both to apprehend relevant features of the presented situation and to know what, in the light of those apprehended features, will fit or fail to fit the rule.[13]

But he has left something out. Correctly applying a rule to a new case will involve a third success: the ability to implement the (cognized) rule in action. Thus, a cognitive explanation of a subject's ability to act rationally would require attributing to the subject grasp of a rule which she implements in action based on her recognition of what the rule requires in a particular situation. It will obviously involve other, related kinds of successes as well—more on this in a moment.

But if my implementation of the principle of continence, say, is needed to move me from an all things considered judgment to action, then why is a higher-order principle of continence not needed to tell me how I am to implement the principle of continence nonakratically, and so on?

To see more specifically how this problem arises, consider the proposal that we attribute the principle as the content of a judgment. Now

---

12. See, for example, Lewis Carroll, "What the Tortoise Said to Achilles," *Mind* 4 (1895): 278–280. Ludwig Wittgenstein, *Philosophical Investigations,* G. E. M. Anscombe, trans. (Oxford: Basil Blackwell, 1953); and Wilfrid Sellars, "Some Reflections on Language Games," *Science, Perception, and Reality* (London: Routledge & Kegan Paul, 1963).

13. Crispin Wright, "Wittgenstein's Rule-Following Considerations and the Central Project of Theoretical Linguistics," (1989), *Rails to Infinity: Essays on Themes from Wittgenstein's* Philosophical Investigations (Cambridge, MA: Harvard University Press, 2001), 170–213.

consider what the status of such a judgment should be; what role would it play in mediating between other mental states and actions? We can try to answer this by borrowing from the Davidsonian model sketched in the first section. Is my holding the principle of continence, for example, tantamount to my having a pro-attitude toward my acting in accordance with my all things considered judgment? This is suggested when Davidson calls the principle of continence a second-order principle: like other principles (and values) it might express a prima facie pro-attitude although unlike other principles, this one is second-order because it would express a pro-attitude about how the deliberation process itself ought to proceed (viz., that the first-order all things considered judgment ought to issue in consistent first-order intentions).[14] If it is to be construed in this way, then we can reasonably ask about the relation between this pro-attitude and any subsequent implementation of the principle. It cannot be a determinate relation prescribed by logic, since the most I can conclude from the principle-qua-pro-attitude is: "Any action which is the formation of a first-order intention is desirable insofar as it accords with my all things considered judgment." But we are still left with the same logical gap between this (second-order) judgment and the kind of all-out judgment needed to motivate the second-order action (in this case, the formation of an all-out first-order intention that the principle of continence was invoked to provide). And in order to fill in this gap we would need something like a third-order principle of continence, and so forth.

Perhaps the principle of continence is the content of an all things considered judgment. Then I judge, all things considered, that I ought to act in accordance with my all things considered judgments. Now the internal regress is explicit. If holding the principle (judging that I ought to act in accordance with my all things considered judgment) were

14. Davidson's comments about the agent not having a *reason* for acting against his all things considered judgment also sustain the interpretation of the principle of continence as a second-order prima facie principle. His comment here implies that such a reason is possible. So, a second-order deliberation process might be set up for deciding whether or not to act on the principle of continence in the first-order case. This would give the (second-order) principle the status of a prima facie judgment.

explanatory of my rational abilities at all, it would only be if the connection between my all things considered judgments and my actions were presupposed. But this was precisely the connection that the principle was invoked to explain.[15]

It ought to be clear that this is going to be a problem with respect to other abilities besides the ability to implement the result of decision-making in action. Other possible fault lines through the practical reasoning process or the logical structure of deliberation which involve perceptual, conceptual, and judgmental abilities will be affected. For each of the possible fault lines we will presumably need a principle to govern the appropriate nonirrational passage; but each of these principles will in their turn be presupposed in the norm-obeying process or in the logical structure of a norm-obeying explanation. Presupposed, then, in the very type of explanation on offer will be our abilities to make appropriate judgments regarding the role of the norm and its applicability, its scope, its relevance, and its overridingness. This is because these judgments and abilities, in turn, will presumably be subject to errors of their own. That this is so is evident when we realize that irrationality (self-deception) affects thoughts, and judgment-formation as well as action, just as wishful thinking affects perception. The principle of continence, then, will have sister principles governing intentional abilities to cover these other possible fault lines in between, and including, perception and action. But all these abilities are presupposed in the logical structure of rule-following explanations: from grasping the rule and circumstances to which it applies, through weighting conflicting principles, interests, plans, and desires, down to intending to implement the rule in action and finally acting in accordance with this intention and implementing the rule in action. All of these abilities are rational abilities, they themselves admit of errors and are thus

15. Suppose, on the other hand, that one tries construing the principle as represented in a second-order standing intention expressing the pro-attitude that I ought to act in accordance with my first-order all things considered judgment, whose adoption necessitates implementation of the principle in action. This move corresponds with conceiving the principle as implemented in causal/psychological processes, which will be discussed in more detail in section 5.

themselves subject to norms which govern their use. But not just some norms or other. Precisely the same norms that are being considered. Thus none of the principles which govern abilities that are presupposed in norm-obeying can be attributed to individuals as objects of thought in order to provide a psychological (norm-obeying) explanation of the disposition to act in accordance with them.[16]

Earlier I suggested that attributing to me knowledge of the laws of physics might explain my ability to answer correctly a question about why I do not float away—even if it cannot explain why I do not float away. Perhaps the analogy can be extended to the norms of rationality. Would attributing to me knowledge of a norm explain my second-order ability to justify my actions that accord with it and diagnose and correct my actions that do not?

But my ability to justify my actions is an ability which itself is, unfortunately, subject to precisely the same threat of irrationality: it is within my justifications that self-deception is evinced. But if these second-order abilities consist in, among other things, a disposition to act in accordance with rational norms in implementing the second-order rules governing justification, then attributing to me knowledge of the norms could not explain my ability to justify, correct, and guide my actions in the light of them for the same reason as before: because the type of explanation proposed presupposes the disposition which is part of that which is to be explained.

---

16. It should be clear that rule-following explanations are not being rejected, tout court. Attributing rules that govern a practice as representations may be explanatory of a person's ability to participate in it as long as certain abilities (namely the ones that are presupposed by rule-following explanations) are not themselves part of the phenomena to be explained by them. Thus my following the rules of chess might be explanatory of abilities I have to make allowable moves in the game, but in this case, the perceptual, conceptual, and intentional abilities which constitute the ability to follow a rule are not the abilities to be explained. If this line of reasoning is correct then anyone interested in giving a cognitive/psychological rule-following type explanation must presuppose that we are rational agents in order to attribute principles which would explain a person's actions; it must be presupposed that we act in accordance with our deliberations (in this case that we not only grasp what the rule requires of us but also implement this understanding in action).

5. Our paradigm example of norm-obeying behavior is when the individual is able to refer to a norm—perhaps she names or describes it—in guiding her deliberations, in justifying her actions, or in diagnosing what went wrong. Perhaps it is this (partly linguistic) ability that would tempt us to attribute to her the norm as an explicit representation. But, as I have urged, *correcting, guiding,* and *justifying* are all themselves rational abilities and presuppose the very same dispositions, and so cannot explain them. This is all very clear in the so-called "self-conscious" cases.

But what if the norm is a *tacit* representation? Might it figure as part of a cognitive explanation nonetheless? Even if the agent cannot self-consciously pick out that stretch of reasoning that the norm governs and self-consciously guide, justify, or correct the transition in the light of her conception of the norm, we might, someone might argue, have evidence that she has got the requisite second-order explicational abilities even if she is not self-conscious of them. And her tacit knowledge of the norms might explain these (sub-cognitive) abilities without threat of regress. Whether or not obtaining the requisite evidence is possible is a question I shall put to one side. Nevertheless, supposing it is possible, unless we had evidence that the agent *conceptualizes* the norm and evidence that it is in virtue of this conceptualization that she acts in accordance with it, we would not be able to distinguish between her merely acting in accordance with the norm and her having the abilities which would be required for (presumptively) attributing tacit knowledge of the norm to her.

But even if obtaining the requisite evidence for the second-order abilities were possible, it should be evident that the same regress problems arise. Whether or not the subject can refer to the norm in her self-conscious explications, if we are to be granted license to attribute (even tacit) knowledge of it to her, her behavior manifesting these second-order sub-cognitive abilities needs to be sufficiently robust. These abilities will involve "sub-perceiving" the norm, "sub-perceiving" the stretch of behavior to which it applies, "sub-judging" its applicability, and "sub-acting" in the light of it. Nevertheless, tacitly to "sub-perceive," "sub-judge," and "sub-implement" it *correctly* as opposed to *irrationally,* presupposes the very same dispositions, operating sub-cognitively, that these tacit representations are, on the hypothesis we are considering, posited to explain.

There is another ostensible way of effecting the appropriate internal connection which might be thought to avoid the regress difficulties. That is, if the second-order justificatory or explicatory abilities which involve the individual's ability to discriminate the norm and guide action in the light of it are somehow "implicitly realized" or "causally determined" in the individual to whom "knowledge" of the norm is attributed.[17]

But now we face a dilemma: either what is implicitly realized is the subject's *cognitive grasp of a norm* or it is not. That is, either the subject manifests abilities which are complex enough presumptively to warrant attributing cognition of the norm to her (or one of her subsystems); in which case this example devolves into one of tacit representation of the principle. And, again, a tacit representation of the principle cannot explain norm-conforming behavior since the disposition to behave in conformity with the norms is presupposed in attributing to the individual the tacit representation, since she or her subsystems have to perceive and implement the norm *correctly:* that is, in conformity with the very same norm of rationality.

Or the subject does not manifest such complex abilities; in which case the example devolves into one in which she merely acts in accordance with the norms. But we would have no more reason to attribute the norm to her as part of her cognitive "hardware" to explain the fact that she acts in accordance with it than we have reason to attribute laws of physics to her cognitive hardware to explain the fact she acts in accordance with them. We might describe her as acting in a way that is subsumed by laws—but in doing so we are making no hypothesis about her psychological or cognitive processes. And this gets at the

17. Although many have argued that it is not really knowledge in this case, since it cannot be considered an achievement, nor for that matter, can it explain a cognitive ability, the cognitive sciences are full of pleas to allow usage of "knowledge" to slacken and to allow the notion of "rule-following" to slacken too, on the grounds that cognitivists are using it with such success. But it is precisely my aim to begin to question whether their use of it is necessary.

intuitive difference between norms and laws. Norms, unlike laws, are *prescriptive* and it is this fact that tempts one to attribute them to the individual as cognitive states or representations in the first place, since doing so at least seems to explain the role they play in *guiding* the individual to whom they are attributed. But attributing them as causal determinants of behavior preempts this function. And this brings us to a correlative problem of conceiving the norms as instantiated in, or as causal determinants of, behavior. Just as their role as prescriptions for guiding behavior would be lost, so, too, would their role in diagnosing rational error, or motivated irrationality. But the norm was introduced to characterize how the individual who has violated it has erred.

It seems that the principle of continence cannot discharge two functions simultaneously: it cannot both be used as diagnostic of irrationality and as that which is causally instantiated in, or determines behavior. Notice that here we run up against the paradox of irrationality as described earlier—and this is a point which prescinds from specific issues about cognition: if the right kinds of relations between mental properties or states are constitutive in this strong sense—if it is necessary that intentional phenomena are either logically or causally determined by norms of rationality that govern them—then irrationality cannot be diagnosed by adverting to the lack of such a relation. In acting akratically, for example, a person has nonetheless acted for reasons and in that sense has acted intentionally. Thus, the sense in which the akratic action is irrational cannot be described as its failure to issue via a relation which is at the same time held to be *necessary* for intentional action.

The point obtains for any proposed principle of rationality: if acting in accordance with it is considered necessary for an action's being intentional, or if its implementation is necessary for a thought's having content, then failing to act in accordance with it, or its failure to be implemented, will not be a possible characterization of an irrational action or thought.[18]

---

18. It is this argument that threatens Davidson's causal account of intentional action. It also suggests a problem with any account that attempts to realize normative relations in

These arguments have important ramifications. There is a strong intuition that we need to make out an *internal* connection between norms and the individual who acts in accordance with them in order to make sense of the intuition that she acts *because* of the norms. A disposition to act in accordance with the norms does not seem to give us the right kind of noncontingent relation required for explanation. But, I argue, this relation cannot be made out as a *cognitive* one such that the norms themselves are objects of knowledge or desired ends and a person engages in reasoning to implement or satisfy them. This is because the "reasoning" here will presuppose the dispositions that attributing these very norms was meant to explain. It would seem, then, that if my disposition to act in conformity with the norms of rationality is indeed some kind of achievement, it is not a *cognitive* one.

6. WHAT OF THE intuitively plausible idea, then, that the norms of rationality *may be* possible objects of thought? Perhaps attributing to me knowledge of a norm of rationality does not *explain* my rational abilities either directly, or via second-order explicational abilities, by the arguments above; but perhaps my having knowledge of the norms *consists* in my ability to justify my actions. And perhaps my having this second-order ability is necessary for me to be considered truly rational. If so, maybe we can make out the sought after "internal" connection after all. My *follow*ing a rule or *obey*ing a norm, as opposed to my merely acting in accordance with it, might *consist* in my ability to justify my actions in the light of the principle prescribing it. And perhaps I can only be considered to be a truly rational agent if I am able to *follow* the rules of rationality in this sense. Correspondingly, I am only akratic when I have the ability to *correct* my actions in the light of the perceived violation of the norm.

But what could it mean to *justify* an action in the light of a norm of rationality? We can make sense, in general, of justifying an action by citing a rule: I might, for example, justify a certain move in chess by citing the rule which allows it. But it would be odd to say that I can

causally necessary conditions for action, thought, or meaning.

justify a rational action (against the incursion of irrationality) by citing a norm of rationality, since the abilities that I am using in the second-order action are precisely the ones with respect to which my first-order action is allegedly being justified. This is not so for the chess case. Someone might ask me to defend my move, and subsequently might ask me to defend my interpretation of the rule that I cite in justification of it, but in so defending my interpretation, *I am not making a move in chess.* But in the case of rationality, I would be making the same kind of move both times. And if my first-order rational dispositions need to be defensible by me in order for me to be considered truly rational, why are my second-order dispositions not in need of the same justification? They would involve, again, the very same moves that allegedly need to be justified in the first place.

Note, finally, a move which is *not* open to those insisting that the ability to *follow* a rule of rationality is necessary for true agency. They are not allowed to appeal to the way our practice does function to argue that—whether or not justification runs out eventually—citing a principle of rationality is what we do, *in fact,* count as justification.[19] Because we do not in fact advert to these principles nor require that anybody do so. Although we learn to cite justifications for all kinds of actions, in so doing our rationality is presupposed.

7. What general conclusions can be gleaned from the arguments above? They suggest that the strategy of construing norms of rationality as objects of cognition, rules to be followed, or norms to be obeyed, is misguided. I argued in the first part of this essay, using Davidson's principle of continence as an example, that individualizing the norm is not necessary for characterizing a subject's thoughts or actions as irrational by her own lights: whether or not attributing the norm to an individual is explanatory or constitutive of her explicational abilities, what would matter for characterizing her as irrational in the light of her own standards is that she act against the norm. But if she

19. This move is arguably available to those who want my following modus ponens, say, on a logic exam to constitute my ability to get the answers right.

has all the abilities required for us to diagnose her as acting against it in the first place, the fact that she acts against it is sufficient for her to be violating a norm which is in some sense her own. If the principles of rationality were to play a normative role for the individual then her inability to express awareness of these norms in explicating her behavior would itself be a sort of failure. But not one which would neutralize an akratic action.

Do they play such a normative role? In the second part of the essay I argued that construing the principle of continence as represented in a psychological or cognitive state of an individual took us nowhere in explaining her disposition to act rationally: either directly or via possible second-order explicational abilities. Internal regresses thwarted attempts to conceive of the principle of continence both as explicitly and tacitly represented. I argued also that this result will threaten norms of a similar ilk: namely those governing perceptual, conceptual, and judgmental abilities that are presupposed in cognitive, rule-following explanations. And although the regress difficulties might be obviated by attempts to construe the "instantiation" of the principle as a causally necessary condition for intentional action, doing so renders the norms explanatorily inert. If we need them at all, we need them as diagnostic of irrationality. But if they are to be construed as causally instantiated in behavior, they have metamorphized into laws and their functions as norms have disappeared.

Indeed, it is the attempt to conceive of the instantiation of a norm of rationality as either a causally or logically necessary condition of intentional action that will render irrational action paradoxical, since if the instantiation of a particular norm is a necessary condition for intentional action or for thought, then irrationality cannot be described as the failure of the instantiation of that norm.

Finally I considered the suggestion that attributing knowledge of the principle to an individual is necessary to explain her ability to justify her action against the incursion of irrationality in the light of the principle. But this is not justification in any nonstipulative and nonartificial sense, since the abilities that she is using in the second-order action are precisely the ones with respect to which her first-order action is allegedly being justified. But in what sense, then, do the norms of rationality govern thought and action if they are not properly construed as objects of cognition? The answer is that they set up the practice of ascribing thoughts and action. This is a point often made by Davidson in discussions about the principle of charity. The principles of rationality seem to play the same kind of role. They are not rules or norms that figure in our attributive practices. They are presupposed by it. But if they ground the practice of interpretation, it would be a category mistake to explain features of the practice by individualizing them.

CHAPTER TWO

Normativity and Thought

1. In "Empiricism and the Philosophy of Mind," Wilfrid Sellars claims that the attempt to analyze epistemic facts in terms of nonepistemic facts is "a radical mistake—a mistake of a piece with the so-called 'naturalistic fallacy' in ethics."[1] I suspect that the attempt to reduce the phenomena or otherwise account for epistemic episodes or states in terms of nonepistemic ones would also be a mistake, and that, generally speaking, the attempt to account for conceptual, intentional, or semantic phenomena in terms of nonconceptual, nonintentional, or nonsemantic phenomena would be instances of a similar error. For Sellars, what distinguishes cognitive episodes, or conceptual awareness, for example, from the having of impressions or sensations (which, properly understood, are states of the perceiver) is the former's role in the "framework of thoughts."[2] In characterizing a state in cognitive or epistemic terms, he says, ". . . we are not giving an empirical description of that episode or state, we are placing it in the

1. Wilfred Sellars, "Empiricism and the Philosophy of Mind," (1956), with an Introduction by Richard Rorty and a Study Guide by Robert Brandom, in *Empiricism and the Philosophy of Mind* (Cambridge, MA: University of Harvard Press, 1997), 19.
2. Ibid., 110.

logical space of reasons, of justifying and being able to justify what one says."[3]

The thought is echoed by a number of subsequent writers: notably by Davidson in his arguments that rationality has no echo in physical theory,[4] by McDowell who claims that "the structure of the space of reasons stubbornly resists being appropriated within a naturalism that conceives nature as the realm of law,"[5] and by Brandom in his program for a normative pragmatics.[6] The nature of the radical mistake that is supposedly involved in accounting for epistemic, conceptual, intentional, semantic, or for short, "normative" phenomena in nonnormative terms is, however, notoriously obscure.

2. David Papineau denies that the normativity of conceptual thought or judgment poses a problem for naturalism. He wishes to defend the idea that norms of judgment are "derived from moral or personal values and do not involve any sui generis species of conceptual normativity."[7] The force of prescriptions such as "in order to achieve truth, you ought to judge in such-and-such ways" derives, he maintains, from "independent moral or personal reasons for attaching value to truth." Papineau's strategy in defending this idea involves assuming a naturalized view of (truth-conditional) content (one which does not itself presuppose norms and one in which the notion of truth itself is construed as a purely "descriptive" property). The strategy also involves pointing to obstacles for accounts that locate normativity "inside" content.

The idea that there could be a "natural" theory of content is, however, precisely what should be challenged by philosophers who are impressed

---

3. Ibid., 76.

4. See especially Donald Davidson, "Mental Events," (1970), and "Psychology as Philosophy," (1974), *Essays on Actions and Events* (Oxford: Clarendon Press, 1980), 207–224 and 229–238.

5. John McDowell, *Mind and World* (Cambridge, MA: Harvard University Press, 1994), 73.

6. Robert Brandom, *Making It Explicit* (Cambridge, MA: Harvard University Press, 1994).

7. David Papineau, "Normativity and Judgment," *Aristotelian Society Supp.* 73, 1 (1999): 17, 18.

with the normativity of thought and its ostensible recalcitrance vis-à-vis the "realm of law."[8] They should not be willing to concede that the "real norms" are ones that can be located "outside" a theory of content. But they should also share Papineau's pessimism about the availability of noncircular theories of norm-governed content. No theory of content (reductive or nonreductive) can be given if this is supposed to involve fashioning the norms that (in some sense) govern thought or interpretation into noncircular necessary and sufficient conditions for the application of mental concepts.

The idea that "real" norms are ones that can be located "outside" content should be resisted because it misidentifies part of the normative space that opens up when an item is identified in cognitive or epistemic terms. This can be seen by considering the aspect of Papineau's proposal that invites us to construe "real" norms as hypothetical imperatives whose prescriptive force is conditional on a believer's desires.[9] There are general questions whether reason-explanations (and in particular, those relevant to an action's moral evaluation) need a conative or motivational element in addition to a cognitive one in order to be truly explanatory. But this familiar dispute about hypothetical vs. categorical prescriptions cannot even get off the ground when the content of the norms with which the agent is in supposed cognitive touch are the alleged norms that govern truth or rationality. Whatever there is

8. Papineau has anticipated this response and has attempted to structure his argument so that the move is question-begging. He claims that his defense of the conditional claim ("*If* you have a naturalist theory of content, *then* you can explain norms of judgment as derived prescriptions orientated to the end of truth") "blocks the argument that naturalist theories of content are inadequate because they cannot account for the normativity of judgment" (ibid., 22). He agrees that theories that locate norms "inside" content call for a species of judgmental norms that resist appropriation within the realm of law. But he rejects the existence of such special content-constituting norms, since he hopes to be able to explain content without any reference to such prior norms in a theory that locates them "outside" content. Technically, then, I agree with Papineau's conditional thesis, but for disappointing reasons: because the antecedent is false. The burden of my discussion in this section is to show why it is not open to Papineau to regard the content-constituting norms as artifacts of the wrong theory of content.

9. The following arguments also tell against the thought that norms are categorical imperatives (which derive from the ostensible moral value of truth).

to the idea that conceptual space is a norm-governed space or, more specifically, to the thought about the normativity of beliefs crystallized in the phrase "beliefs aim at the truth,"[10] this cannot be construed as a command issued to persons to conform or mold their judgments to meet the requirements of truth, or as giving them a reason (categorical or hypothetical) to form only true beliefs.

Belief-acquisition is not an action. Although one can form strategies (for example, by avoiding relevant evidence) that will maximize or minimize one's chances of acquiring some particular belief, forming beliefs is not something one chooses to do. Furthermore, the acquisition of true or false beliefs, even if this were construed as a quasi action, could not possibly be explained by attributing to an individual a desire or an aversion to conform to a norm. In order for such an explanation to get off the ground, the individual would have to have beliefs about the norm—how else could it be the object of her desires?—and presumably questions regarding the truth of her beliefs about the norm would simply resurface. It ought to be clear in any case that the explanatory mileage gained by attributing to an individual the desire to seek the truth would be negligible, at best (even if this does not involve putting her in cognitive contact with a norm). The desire for truth can only explain one's decision to stop visiting the astrologer if one comes to believe, for example, that the interpretation of horoscopes is not a reliable process for gleaning the truth. But how is the truth-aspiration of the belief about the unreliability of astrology supposed to be understood? It is presupposed.[11]

Contentful states cannot play a role in reason-explanation unless it is taken for granted that beliefs "aim at the truth." The contingent desires of the agent have nothing to do with this supposition: it is a conceptual truth about beliefs. There are, of course, special circumstances in

---

10. Bernard Williams, "Deciding to Believe," *Problems of the Self* (Cambridge: Cambridge University Press, 1973), 136–151.

11. Although I have framed this as a problem inherent in the idea that someone might be said to obey or to want to accord with semantic norms, the problem easily generalizes to include the attempt to construe rationality as a matter of obeying norms of rationality. See "De-Individualizing Norms of Rationality" (chapter 1).

which it makes sense to talk about a person who wants the truth, or who wants to avoid it, just as there are special circumstances in which it makes sense to talk about a person who might need to be convinced that some of her belief-forming processes are unreliable. That these are *special* circumstances can be seen by reflecting on the fact that "the game of giving and asking for reasons," the practice of justification, and, incidentally, the whole notion of *explanation* presupposes truth both as an ideal and as the norm. The framework of thought, or the logical space of reasons, comes "prepackaged" with truth and its preservation as standards. Although instances can be isolated in which a person fails to conform to modus ponens, or seeks to avoid the truth, or acts against her better judgment, this is accomplished against a general background in which truth, inference, and rationality remain intact.[12]

At least part of the normativity that constitutes the "logical space of reason" attaches to this *background* space; it attaches to what is presupposed when items are identified in conceptual space, or in cognitive or intentional terms. The attempt to locate the "norms of thought" or the "norms of judgment" as grounded in a person's desires or reasons therefore mislocates part of the normativity that is said to resist appropriation within a world that conceives nature as the realm of law.

3. I HAVE ARGUED that Papineau misconstrues the normativity that attaches to what Sellars calls the "framework of thought" or the "logical space of reasons," and I have suggested that the right way to conceive it is in terms of the relations that are taken for granted in reason-explanation. I have also hinted at the futility of construing the norms that are extracted by studying these relations as playing a reason-giving role in

12. This should not be taken to imply that the background space (or any one of its features) is itself something that cannot be put in jeopardy; but this could only be done once the framework for dismantling it no longer presupposes it. Compare Sellars's claim that ". . . empirical knowledge, like its sophisticated extension, science, is rational, not because it has a *foundation* but because it is a self-correcting enterprise which can put *any* claim in jeopardy, though not *all* at once." *Empiricism and the Philosophy of Mind,* 79. Correlatively, the fact that this framework is presupposed should not be taken to imply that it is necessary: human patterns of thought might have been very different. See Barry Stroud, "Wittgenstein and Logical Necessity," *Philosophical Review* 74 (1965): 504–518.

an explanation of an individual's rationality. In other words, attributing to an individual beliefs about, or the desire to conform to, a norm of rationality cannot explain—on pain of regress—her ability to act rationally or her ability to form true beliefs: these abilities are presupposed in the very idea that she has beliefs and desires. This last thought is related to the pessimism I endorsed about the availability of noncircular accounts of items identified within the space of reasons, since it seems that the phenomena to be explained are presupposed by the kind of explanation on offer. I shall come back to this thought later. First, more needs to be said about why in particular it is a mistake to attempt to reduce items within the "framework of thought" to the realm of law.

One way of approaching the question would be to see what is missing when such a reduction is attempted. Calibrating something against a norm, ideal, or standard allows it to be evaluated in terms of its relative success or failure in achieving this ideal or standard. The appeal to semantic norms (e.g., that "*vert(e)*" means "green") or to principles of rationality (e.g., that a person tends to act in accordance with her better judgment) allows a diagnosis of, for example, what has gone wrong in a particular case when a language user misunderstands a word or when a rational agent acts irrationally. Items identified within a normative space admit of this special kind of error—"internal" error—which allows both that the one who commits the error is still participating in the practice that the norms govern and that her mistaken moves are still moves within that practice. Someone who makes mistakes in inference is still a thinker; someone who mistakes the meaning of a word might still say something meaningful; someone who acts irrationally is still acting in a way that is rationalizable; and someone who believes something falsely still has a belief. The phenomena to be explained—thinking, acting, saying, believing—allow "internal" mistakes of this sort: mistakes, like an *error* in baseball, that do not disqualify the player and are still considered to be moves within the game.

Various attempts to assimilate normative phenomena to law-governed events have run into difficulty accommodating this kind of error. In contemporary philosophy of mind, the difficulty shows up in the computationalist's attempt to construe the algorithmic/heuristic rules that govern transitions between representations as themselves instantiated causally (in the neuronal structure of the brain), and it shows up in the representationalist's attempt to give causally necessary and sufficient conditions of what it is to be a content-bearing state.[13] It also shows up in certain functionalists' construal of mental terms as theoretical posits in a causal/nomological "theory of mind."[14] But the difficulty accommodating "internal error" is not merely a problem for naturalists: it is a problem for anyone who misunderstands the nature of the platitudes or principles that are presupposed with the application of concepts that have their home in the "logical space of reasons." It is a difficulty, that is, for any philosophical account of the mental that misunderstands the kernel of truth contained in the idea, for example, that it is "constitutive of [one's] possessing a belief with a certain truth condition that [one] be subject to norms which are apt to guide [one] to the truth."[15] If the "instantiation" of such a norm (as it were) is conceived *either as a causally or a logically necessary condition* of having a belief, being an action, or being a perception, then false beliefs, irrational actions, and misperceptions will be impossible. Take any norm that allegedly governs, say, intentional action: if the instantiation of that norm is a necessary condition for intentional action, then irrational actions cannot be described as the failure of the instantiation of that norm. This idea gets to the heart of the philosophical paradox of irrationally,[16] and the way out is to recognize that the norms that govern a practice do not provide rigid conditions of what is to count as a move within that practice. The appeal to norms supports the distinction between successful moves and unsuccessful ones: true and false beliefs; misperceptions and veridical ones; rational actions and irrational ones.

The idea that "psychological principles" can be extracted from an investigation of our reason-giving practices is nonetheless attractive. (As is the idea that there are underlying principles which might be

13. See "How to Resist Mental Representations" (chapter 9), "Conceptual Analysis, Theory Construction, and Philosophical Elucidation in the Philosophy of Mind" (chapter 11), and "Ryle's Regress and Cognitive Science" (chapter 12) for further discussion.

14. See "Why Reasons May Not Be Causes" (chapter 5).

15. Papineau, "Normativity and Judgment," 24.

16. See "De-Individualizing Norms of Rationality" (chapter 1).

extracted by looking at our ethical practices.) Suppose we met with some success in extracting such principles or norms by watching how people correct themselves and others and how they justify these corrections. The list of generalizations that results might form a kind of normative guide for participating in the practice: a list of dos and don'ts. And knowledge of the norms might give us some understanding of the nature of the practice. But the sense in which this normative guide forms a *theory* would probably turn out to be similar to the sense in which adumbrating the dos and don'ts of etiquette or classical piano-playing forms a theory. One would not think that the norms are exhaustive, that they suffice to govern allowable moves, that they are exceptionless, or that they are even stable. But this is a very different conception from that which is involved in discovering natural laws.

4. A CONTROVERSIAL QUESTION (even among those sympathetic with this general line of thinking) is whether one can view the nature of our reason-giving practices independently of the choices, decisions, and reactions of the participants; the answer would be "no" according to those who are sympathetic with the idea that the choices, decisions, and reactions may be partially constitutive of the protean phenomena under investigation. This is certainly the case for etiquette and piano-playing, and plausibly the case for aesthetics and ethics.[17] If this intuition were true for our reason-giving practices as well, it would pose another threat to a "naturalist" program which construed its investigation of the mental along the lines of a scientific model, since the whole point of a scientific investigation is to prescind from subjective features as much as possible.

I suggested earlier that at least part of the normativity that constitutes the "logical space of reason" attaches to what is presupposed when we identify something as an item in conceptual space or in cognitive or intentional terms. The normativity attaches to this background space,

17. I note here that I am passing quickly over an important and interesting distinction between individual choices, decisions, and reactions, and group choices, decisions, and reactions.

I argued, quite independently of considerations having to do with particular reasons of particular individuals. But I think that another part of the normativity that attaches to the logical space of reasons will concern particular individuals and their "points of view."

In an earlier work, Papineau suggests that the only serious candidate for an argument that there are certain principles of thought necessarily common to all thinkers is the argument that "the interpretation of human thought is a distinctive matter of *understanding* (or *verstehen* . . .)."[18] He glosses this in terms of the familiar argument that if what is involved in making sense of a person amounts to appreciating how certain beliefs, judgments, or actions might be rational, given the other things the subject believes, desires, and so forth, then principles of rationality must be in play in the very identification of an item as one with content. But Papineau suggests that "the whole idea of *verstehen* as a distinctive mode of understanding loses its appeal once we reject the idea of incorrigible private access to mental 'givens.'"[19]

Now it would indeed be ironic if it turned out that any part of the special understanding that is available through the identification of items within what Sellars calls the "space of reasons" depends upon accepting the very myth that Sellars deserves so much of the credit for repudiating. In any case, it is difficult to see why the special mode of understanding that is generated by attributing reasons *should* depend upon accepting some version of the myth of the given.

Sellars's attack on the myth of the given is an attack on the empiricist's appeal to various items in order to justify, logically ground, or provide a foundation for claims to empirical knowledge. These items include "immediate experiences," "self-authenticating episodes," "knowings in presence," objects of "direct acquaintance" or other "private" episodes, and indeed, "immediate observables" within the public domain or even facts, if the capacity to know these facts is supposed to figure in an explanation of the acquisition of ordinary empirical concepts. Sellars famously allows for the existence of inner episodes, immediate

18. David Papineau, *Reality and Representation* (Oxford: Basil Blackwell, 1987), 119.
19. Ibid., 121.

experiences, or appearings, as long as it is understood that these cannot be the raw data for the application of concepts; according to him, we can only make sense of *looks-talk* as parasitic on and thus presupposing (in the order of analysis) *sees-talk* or awareness of ordinary objects. To the extent that there are nonverbal thoughts, these cannot be what (in the order of analysis) accounts for language: the intentionality of thoughts is, according to Sellars, inherited from and hence parasitic upon the semantical categories pertaining to overt verbal performances.

The sense-data theorist is an example of someone who makes the category mistake of supposing that the having of a sensation, which is not supposed to tax conceptual resources and is available to brutes and language users alike, is magically supposed to carry with it the inferential structure (or "articulation") which would enable it to play a justificatory role (i.e., be "usable in formulating premises and conclusions of inferences assessable as correct or incorrect"[20]). The mistake, fallacy, or myth is to suppose "that the space of reasons, the space of justifications or warrants, extends more widely than the conceptual sphere" [21] or that "what is Given can serve as justification, without its being given requiring the exercise of conceptual capacities"[22] or that anything "can count as a reason for a belief except another belief."[23]

It is true that the kind of understanding that is brought about by identifying an item as something that falls within the space of reasons is very different from the kind of understanding that is brought about by identifying an item as something that falls within a causal nexus; presumably, it is this fundamental difference that accounts for the suspicion that there is no recipe (via analysis or reduction) for identifying an item identified within one space as an item within the other. Papineau is right in supposing that some of what might account for the

20. Brandom, in his commentary in Sellars, *Empiricism and the Philosophy of Mind,* n38.
21. McDowell, *Mind and World,* 7, quoted in Brandom, ibid., n122.
22. Brandom, ibid., 122.
23. Donald Davidson, "A Coherence Theory of Truth and Knowledge," (1983), *Subjective, Intersubjective, Objective* (Oxford: Clarendon Press, 2001), 137–153. This is quoted and discussed in McDowell, *Mind and World,* especially 14ff and 137ff, and in Brandom, ibid., 122.

difference will take into account the point of view of the person whose thoughts and actions are under scrutiny. But a point of view in this sense does not introduce particular episodes of consciousness that are supposed to be independent of our conceptual resources and provide the sort of foundation sought by empiricists for claims to knowledge.

There is an obvious sense in which, when we ascribe a belief to someone as part of a way of rationalizing *her* actions and the rest of *her* mental life, we are bound to be considering things from *her* point of view. But there is a different way of coming to see the importance of the individual's point of view which I would like to consider—one I think Sellars is getting at when he suggests (in the passage cited in section 1) that the logical space of reasons concerns "justifying and being able to justify what one says." I have in mind the distinction that we often need to make when we are considering whether to regard someone as a fully fledged participant in some practice or not. Sometimes, for example, we need to be able to distinguish between an individual who is able to make a move in a game, accidentally, as it were, and an individual who makes a move in a game because it is a move in the game.[24] The distinction is familiar enough when it comes to deciding whom to accept and whom to reject as players in the "game of giving and asking for reasons." We tend to reject without difficulty inanimate objects, forms of plant life, simple biological creatures and simple mechanical devices, even when, on occasion, they behave in ways that accord with rational norms. But difficulties arise with more complicated biological creatures like cats, dogs, and young children, and with the possibility of artificial intelligence. Papineau, of course, is right to suggest that it makes sense to attribute conceptual awareness to animals and to children, and that it makes sense to ascribe beliefs and desires to them, since doing so enables us to understand their behavior by showing why it would be rational for them to behave in that way. But there are obvious limits to their rationality. Crucially missing in the case of animals and small

24. See "Playing the Rule-Following Game" (chapter 3) and Wilfrid Sellars, "Some Reflections on Language-Games," *Science, Perception, and Reality* (London: Routledge & Kegan Paul, 1963).

children, but present in paradigm cases of agency, is the agent's own explanation of or "take" on her reasons for acting. Now the importance of the agent's self-conception has long been discounted—both by those (e.g., following Freud) who are attracted to the idea that a person's beliefs, desires, etc., might be unconscious and therefore not accessible for reporting purposes and by those philosophers of language and mind (e.g., those following Chomsky and Fodor) who invoke a notion of tacitly held beliefs or quasi-intentional informational states which are also not accessible for reporting purposes. Those who discount the paradigm cases in which an agent or a speaker is able to offer reasons for which she acted, or explain the meaning of her words, tend to do so by pointing out that agents and speakers are often wrong about their reasons and meanings and that often a person's sincere explanation of her actions or words needs to be rejected.

It is certainly true that a person's authority about her reasons or meanings does not provide a guarantee that what she says will rationalize her actions or constitute the meaning of her words. Nonetheless, what she says about these things cannot be ignored, even when she is wrong, since what she says is crucial to understanding her. It is not simply that we would be unable to diagnose a person as self-deceived unless we had some access to her own (mistaken) self-understanding. But insofar as an agent's conception of herself sets its own rational constraints on her future actions (and reinterpretations of herself) we would also lose a whole dimension of what is involved when we construe an individual as a participant in the "game of giving and asking for reasons."

It will be worth elaborating these ideas. Consider first cases of self-deception, where it should be noted that there are at least two strands of behavioral patterns manifested. One is the pattern that belies the agent's self-conception and uncovers her ignorance about her own mind. The other is the pattern—often of denials, of protestations, of avoidance—that is a straightforwardly rationalizable outcome of this self-conception. Constant's Adolphe, who is self-deceived about the obstacles to his worldly success, is not merely wrong to blame his relationship with Ellénore. His false conception about their life together feeds into a whole pattern of behavior leading to a tragedy that is itself only rendered comprehensible by this conception.

The importance of a person's self-conception is reflected in other ways besides in cases of systematic self-deception. When I want to hear that my beloved loves me, it is not merely because I like the sound of the words.[25] It is because his own conception of his love for me will account for a much richer pattern of behavior than his unavowed love for me would. Imagine the difference between a relationship in which the beloved's love forms a part of his self-understanding which is expressed comfortably not only in his plans and actions but in his thoughts about these plans—and a relationship in which the beloved's love for the other—though it finds expression in certain ways—does not form a part of his own "narrative."

Most famously, this distinction plays a fundamental role when we attempt to evaluate the moral worth of an action. Kant probably went too far in suggesting that actions that accord with duty but which are not performed from duty have no moral worth.[26] But there is something right about the idea that the capacity to justify or to provide the right sort of reasons for one's action is an integral part of what it is to see someone as a moral agent. There is some plausibility, as well, in the idea that the capacity to justify or to provide reasons for one's actions or for one's judgments is an integral part of what it is to see someone as an agent or as someone who has a claim to knowledge.[27]

---

25. This example was suggested by a point made by Richard Moran in his reply to Bar-On and Long at the American Philosophical Association Pacific Division Meeting in Los Angeles, Spring, 1998.

26. Immanuel Kant, *Foundations of the Metaphysics of Morals,* Lewis White Beck, trans. (Indianapolis, IN: Bobbs-Merrill, 1959), first section.

27. This idea is integral to Sellars's conception of what counts as knowledge. He concedes that ". . . the correctness of a report [say, "This is green"] does not have to be construed as the rightness of an *action*" and that "it can be correct as being an instance of a general mode of behaviour which, in a given linguistic community, it is reasonable to sanction and support." Nonetheless, there is a second strand to the authority of a sentence token "This is green": it must not only have authority in the sense that we can infer from the utterance the existence of a green object, which we could do, by the way, for a well-trained pigeon. What distinguishes a fully paid-up member in the practice (someone we can credit with knowledge) is that "this authority must *in some sense* be recognized by the person whose report it is." *Empiricism and the Philosophy of Mind,* 74. And, as Sellars acknowledges,

It is this capacity to justify that is missing when the individuals whose actions are being evaluated are animals and young children. There is nothing strange about accepting them as fledgling members of our reason-giving (and epistemic) practices if this merely involves identifying their behavior as conforming to a rational pattern—even, in certain cases, one which comes about as a result of training—but the sense in which they are not full members, and might never become them, must not be forgotten. They are simply not responsible to the norms of rationality even if, in some sense, they can be seen as responsive to them.[28]

5. Normative phenomena cannot be converted into nonnormative phenomena without the loss of the special explanatory apparatus that the appeal to norms makes possible. I have suggested that this will involve a dimension of assessment which allows that a mistaken move can still be a move within the practice and that an individual who makes the mistake can still be considered a participant in the practice. I have also suggested that it will involve (not unconnectedly) the individual's own point of view or self-conception. Science, conceived as the realm of law, makes no room for "internal error" in the same sense, nor for subjective points of view. The explanatory aims of science are different.

Even if one were to try to give some modified scientific account of this notion of "self-conception" that seems such an important strand

> this is a steep hurdle indeed. For if the authority of the report "This is green" lies in the fact that the existence of green items appropriately related to the perceiver can be inferred from the occurrence of such reports, it follows that only a person who is able to draw this inference, and therefore who has not only the concept *green,* but also the concept of uttering "This is green"—indeed, the concept of certain conditions of perception, those which would correctly be called "standard conditions"—could be in a position to token "This is green" in recognition of its authority. (Ibid., 75)

It follows from this that one could not have observational knowledge of any fact unless one knew many *other* things as well. In the next section, I suggest that the same sort of considerations apply to rational agency.

28. See Michael Root's discussion in "Davidson and Social Science," *Actions and Events: Perspectives on the Philosophy of Donald Davidson,* Ernest Lepore and Brian McLaughlin, eds. (Oxford: Basil Blackwell, 1985), 272–304.

in conceiving an individual as a rational agent, how could this possibly be achieved? The normal way of deciding whether someone is acting in accordance with the rules of a game merely by accident, because of training, or because she is *following* those rules, is to decide whether it makes sense to see her as having a cognitive grasp of the rules. The usual way of deciding this is to determine whether she is capable of appealing to them in an attempt to justify or to explain what she has done. Obviously, however, if the capacity to see oneself as acting for reasons is part of what it is to be a rational agent (or if the capacity to justify one's beliefs is part of what it is to be a full member of our interpretive practices), then no noncircular account or explanation of what it is to act rationally, have reasons, or beliefs can be given. To paraphrase Sellars,

> . . . while the process of becoming a rational agent may—indeed does—involve a long history of acquiring *piecemeal* habits of response to behavioral patterns, there is an important sense in which one has *not yet committed* an action unless one is already a rational agent.[29]

This line of thinking, I believe, gives some substance to the refusal to offer constitutive accounts of normative phenomena, or to the refusal to "locate 'bedrock' lower than it is."[30] Papineau, along with many others, is dismissive of those who hold that "the norms of judgment are primitive and not to be further explained."[31] But if I am right, the refusal to offer such an explanation is justified by recognizing the impossibility

29. This is a paraphrase of:

> . . . while the process of acquiring the concept green may—indeed does—involve a long history of acquiring *piecemeal* habits of response to various objects in various circumstances, there is an important sense in which one has *no* concept pertaining to the observable properties of physical objects in Space and Time unless one has them all—and indeed, as we shall see, a great deal more besides. (Sellars, *Empiricism and the Philosophy of Mind,* 45)

30. John McDowell, "Wittgenstein on Following a Rule," (1983), *Mind, Value and Reality* (Cambridge, MA: Harvard University Press, 1998), 252.

31. Papineau, "Normativity and Judgment," 28.

of giving an account of something that is at the same time presupposed in our very method of accounting for it.

I would like to end by lodging a complaint against the use of the term "nonnaturalist" as a description of those who take epistemic or conceptual episodes to be irreducible to particulars that operate within the realm of law. Note that Sellars's dispute with the empiricists was about the incommensurability of items *within different areas of discourse:* the extent to which he believed that "science is the measure of all things"[32] ought to be consistent with his mockery of the idea that there is a way to get beyond all discourse.[33] Note too that Sellars himself has been described as explaining "why we can be naturalists without being behaviourists, why we can accept Wittgenstein's doubts about what Sellars calls 'self-authenticating non-verbal episodes' without sharing Ryle's doubts about the existence of such mental entities as thoughts and sense-impressions"[34] and as giving us a way to understand "mind as gradually entering the universe by and through the gradual development of language, as part of a naturalistically explicable evolutionary process, rather than seeing language as the outward manifestation of something inward and mysterious which humans have and animals lack."[35]

The philosophers who follow Sellars in supposing that normative phenomena are basic are "naturalists" fighting a battle against scientism. One familiar manifestation of scientism is the idea that in order to render the phenomena under investigation understandable or explicable, we have to be able to take a point of view that prescinds as much as possible from human sensibilities, and we have to reconstrue

32. Sellars, *Empiricism and the Philosophy of Mind,* 83.

33. "Does the reader not recognize Jones as Man himself in the middle of his journey from the grunts and groans of the cave to the subtle and polydimensional discourse of the drawing room, the laboratory, and the study, the language of Henry and William James, of Einstein and of the philosophers who, in their efforts to break out of discourse to an *arché* beyond discourse, have provided the most curious dimension of all." Sellars, *Empiricism and the Philosophy of Mind,* 117.

34. Richard Rorty, in the Introduction to Sellars, *Empiricism and the Philosophy of Mind,* 6.

35. Ibid., 7.

areas of discourse in which the application of their proprietary concepts (evaluative, conative, and secondary quality concepts, in particular) seem to be constitutively tied to human sensibilities. It is becoming recognized, however, that "evaluability" or "normativity" extends wider than usually supposed,[36] so it is no longer clear what sense there is in programs designed to reconstrue or "legitimatize" this element.

McDowell has long urged that not only *realists* of a certain breed, but also certain *irrealists* make a mistake in wishing to "legitimatize" or reconstrue domains of discourse by relating them to areas less immune to human sensibility.[37] He might be right that the conceptual is unbounded,[38] that the escape from human sensibilities is a misconceived project in any case, and that there cannot be a project of giving an account of normative discourse if that means doing it from a non-normative point of view. Conceding all of this, it seems to me that there are residual differences between areas of discourse worth exploring that might give some substance (back) to the idea that in certain domains, individual human judgment plays more or less of a role than it does in others.[39] In the domains that involve "giving and asking for reasons," as my remarks above indicate, I suspect the individual's contribution plays a very large role.[40]

---

36. See, for example, Brandom, *Making It Explicit.*

37. See, for example, McDowell, "Non-Cognitivism and Rule-Following," (1981), *Mind, Value and Reality,* 198–218.

38. McDowell, *Mind and World.*

39. Compare Crispin Wright, *Realism, Meaning and Truth,* 2nd edn. (Oxford: Blackwell, 1993).

40. I discuss this further in "Some Constructivist Thoughts About Self-Knowledge" and "Self-Knowledge, Normativity, and Construction" (chapters 13 and 14).

CHAPTER THREE

Playing the Rule-Following Game

> What this shows is that there is a way of grasping a rule which is not an interpretation, but which is exhibited in what we call "obeying a rule" and "going against it" in actual cases.
>
> Wittgenstein, *Philosophical Investigations*, § 201

I. Introduction

1. IT HAS BEEN suggested that in order to make the study of meaning more manageable, we ought to consider what must be known by an individual who is able to understand and speak a language. Philosophical questions about the nature of meaning have thus transformed into questions about the form that should be taken by an idealized theory of meaning, knowledge of which would suffice to explain a speaker's competence. Some scientifically minded linguists, psychologists, and philosophers are attracted to this way of investigating language because they are optimistic that meanings—conceived as the contents expressed by the theorems of a meaning theory—can be understood as abstract representations in the brains of language users. Thus, the study of meaning, as such, is thought by some to invite a cognitive-scientific investigation grounded, optimistically, in the lower reaches of neuroscience. Recently, philosophers and psychologists have sought similar types of explanations of the ability to understand and predict rational behavior by attributing to competent individuals a theory of mind, knowledge of which would explain these abilities, or correlatively, the absence of which would explain why certain individuals lack the ability to understand and predict the behavior of others. Within the last decade, research in the cognitive sciences has focused on experimentation designed to test this hypothesis.

Rules or norms in some sense govern various social practices. An individual's knowledge of the norms is supposed to figure in an explanation of her ability to participate in the practice. But I suspect that there is something wrong—deeply wrong—with the attempt to give rule-following explanations of broadly rational activities, and the problems inherent in such attempts do not seem to be solved by supposing that the rules are expressed in the contents of subcognitive states. Although I shall not be able to argue for it here, I suspect the difficulties in these projects extend to what is taken for granted in much contemporary philosophy of mind, as they threaten certain accounts of reason-explanation and present a challenge to the very idea of causally efficacious, content-bearing states.[1]

In part 2 of this chapter I shall develop a very general argument that raises a prima facie doubt about the coherence of certain attempts by theorists of meaning and mind that involve attributing knowledge of the relevant theory to an individual in order to explain her linguistic or rational abilities. (Henceforth, I call these "cognitive explanations" and refer to the theorist who attempts this kind of explanation as a "cognitivist.") The argument takes the form of a reductio in which the premises the opponent seems bound to accept lead to a dilemma, neither horn of which is viable. On one horn of the dilemma a vicious explanatory regress ensues. On the other horn, the regress may be halted only at the cost of destroying the normative nature of the practice, thus rendering attribution of *knowledge* of the norms pointless. I show how the dilemma arises by focusing on a norm-governed practice for which rule-following explanations make sense (I choose the game of baseball for this purpose). I argue that it is only when our broadly rational abilities are *presupposed* that such an explanation might even get off the ground.

The attempt to abstract these broadly rational abilities and give systematic, cognitive explanations of them leads to the dilemma—a

1. These intuitions are developed in the chapters of part 2 and part 3 of this volume.

dilemma that confronts *any* attempt to give cognitive explanations of abilities whose broadly rational character is part of what needs to be explained. This includes, among others, the ability to understand and speak a language, the ability to understand and predict rational behavior, and the ability to act rationally. I argue that the appeal to implicit knowledge does not manage to sidestep the regress, whereas the appeal to cognitive processes sidesteps it, only to become impaled on the other horn of the dilemma. In part 3, I consider in more detail how this dilemma affects the idea that rational action is itself an ability that admits of rules or norms that govern it. If one supposes that knowledge of such rules is necessary for reasonable action, the same kind of dilemma arises and I show the negative implications of this dilemma for nomological accounts of reason-explanation. In part 4, I consider the general ramifications of this discussion on rule-following for explanations involving the attribution of knowledge.

II. Cognitive Explanations of Rule-Following Abilities

2. In order to get a sense of when rule-following explanations make at least prima facie sense, it will be useful to consider ordinary games. Games are relevant because they are typically governed by rules. The introduction of rules into the analysis of games seems indicated simply because they provide a standard in virtue of which it makes sense to judge particular moves in the game as "correct" or "incorrect."[2] Human practices or activities of various kinds can be assimilated to games insofar as they, too, involve "moves" that are apt for judgments of correctness or incorrectness. Consider the game of baseball. Call the

2. This is a minimal sense in which talk of rules is indicated. A serious topography of the various sorts of rules governing an activity such as baseball would be quite complex. It would involve, among other things, distinguishing regulative and constitutive rules; subrules governing professional games vs. amateur games; National League vs. American League; rules for umpires; rules for compiling statistics on the individual players, etc. My interest, in the main, is in the rules that are candidates for "constitutive" rules—i.e., those that make baseball the particular kind of sports activity it is—and in what the baseball player's relationship to those rules is supposed to be.

moves that are made in the game of baseball—hitting the ball with a bat, running around the bases, etc.,—"material" moves, or moves made in the "object game."[3] The rules of baseball in some sense govern the material moves, but in what sense? Someone who knows nothing about baseball and whom we would be reluctant to describe as playing the game might make some of the material moves "accidentally," as it were. Conversely, someone we would want to describe as playing the game might fail to make appropriate material moves (this person would be playing baseball incorrectly). In order to credit an individual with playing baseball we might seek some kind of "internal connection" between her and the rules so that we can say that she is *making (or attempting) the moves as part of the game of baseball.* We might be satisfied if she were able to play what Sellars calls the "metagame": i.e., if she were able, using language, to cite the rules of baseball, and to make a case for the material moves being sanctioned by the rules.[4] Perhaps here there would be nothing objectionable in describing this internal connection as a type of knowledge: someone who is able to cite the rules of baseball at the appropriate time might be said to *know* the rules, and this knowledge might figure in an explanation of the moves she makes. Perhaps this knowledge might figure in an explanation of her ability to play the game.

3. Suppose we are satisfied that an individual is making a move in the game because we have established the sought-after internal connection—this individual is manifesting *rule-following* as opposed to mere rule-conforming behavior. She is not only able to make the material moves of the game, she is also able to correct or justify her moves by appealing to the rules of the game. But there is still a sense in which this individual, too, in being able to play not only the object game but

---

3. I am borrowing these terms, as well as some of the framework for discussion, from Wilfrid Sellars, "Some Reflections on Language Games," *Science, Perception, and Reality* (London: Routledge & Kegan Paul, 1963).

4. The suggestion is that the ability to cite the rules might be considered a (defeasible) sufficient condition for establishing the sought-after "internal connection." Whether or not it is also a necessary condition will be considered in the last section of this chapter.

the metagame as well, seems to be evincing general abilities—abilities involved in the very activity of following a rule—that might themselves go wrong.

Imagine, for example, someone who is being taught how to play baseball. Suppose she has a general understanding of the language her coach uses to describe the rules, yet she misjudges the domain of one of them: she takes it to apply to a situation to which it does not apply. Or suppose she can convince us of her general ability to apply rules in appropriate situations, but she cannot understand the meaning of some of her coach's words and therefore does not understand the rule she is supposed to be following. Or suppose she understands her coach's language, she understands the type of situations to which the rules apply, but she fails, at times, to act in the light of this understanding. Such errors are not happily described as failures to act in conformity with the rules of baseball for they may occur even if a person makes material moves that happen to conform to them. Rather, such errors (which may involve failure to understand the rule, failure to understand the relevant domain, or failure to implement one's understanding in action) thwart the attempt to follow the rules of baseball. People misfollow rules all the time and as a result tend to make mistakes: they bungle recipes and produce culinary disasters; they misinterpret or misperceive signposts and travel in the wrong direction; and they make mistakes in following orders and are punished.

Following the rules of baseball or following recipes or maps—like making the material moves involved in playing baseball, or in cooking, or in traveling to a particular destination—are things that one can do correctly or incorrectly. This fact invites us to construe the activities that are involved in rule-following as themselves rule-governed, for it seems to be a platitude that error is diagnosable only insofar as a norm or standard has been violated. More concisely, we might say that the various kinds of activities involved in rule-following constitute a game, or a set of games. (By this I mean only that the appeal to rules, norms, or standards seems apt when a diagnosis of error can be made.) It should be clear, however, that if rule-following is itself a kind of a game, it is not regulated by the sort of rules we have been considering—the rules of baseball do not tell us how to follow them. The rule-following game would consist in a different set of games which would require construing the appeal to the rules of baseball, or the appeal to a recipe, as a constituent "move." Indeed, one might think that the abilities that are manifested might be construed quite generally, abstracted from any particular object game, like baseball or cooking. If, for example, the coach's instructions are given in some language, then the abilities necessary for rule-following in this particular case would involve the ability to understand this language. If implementing the coach's instructions involves recognizing the domains in which they apply, then the ability to follow a rule here would include the appropriate conceptual abilities.

The general abilities that are manifested in rule-following are those that are involved in implementing one's understanding of a rule in action, in judgment, or in speech. These will include whatever conceptual, perceptual, interpretive, and linguistic abilities are needed to grasp the rule and to understand the situations in which it applies and whatever inferential and rational abilities are needed in order to act in the light of one's understanding of it. Rule-following abilities include, that is, those abilities that are implicated in our ordinary notions of meaning, understanding, thinking, reasoning, and acting. Now, no one has explicitly attempted to give a cognitive explanation of the ability to follow a rule or play the rule-following game. However, many have attempted to provide systematic theories of language, rational thought, and action; indeed, this has been one of the central tasks of analytic philosophy—to offer, that is, accounts of the general abilities that turn out to be (as I shall argue) constitutive of, or part of what it is involved in, rule-following. Those who attempt to advance cognitive explanations of any of these abilities broadly construed might as well be trying to propose such explanations for the ability to play the rule-following game.

I suspect that there might be a problem in attempting to construe the constituent abilities that are involved in rule-following in a very general sort of way—in a way that abstracts from their role in an object game.[5] I will not pursue this line here. Whether or not this is a prob-

5. See, for example, Ludwig Wittgenstein, *On Certainty*, G. E. M Anscombe and G. H. von Wright, eds. (Oxford: Basil Blackwell, 1960), §§28, 29, and 43–49.

lem, it certainly makes no sense to attribute knowledge of the putative norms that govern these activities to an individual as part of a substantive explanation of her ability to participate in them. It makes sense to explain an individual's material move in baseball (say, the batter's merely tapping the ball with the bat and thus inviting the pitcher to throw her out at first base) by attributing to her knowledge of some or other rules (e.g., about sacrifice strategies). But extending this kind of explanation to the case in which the material moves are the (generally and abstractly construed) constituent abilities involved in rule-following is incoherent. Or so I shall argue.

4. We were motivated to speak of someone's following a rule in the first place in order to discern some kind of internal connection between the individual who makes the moves and the rules that govern them. We were seeking this connection to assure us that the individual whose behavior accords with the rules is a participant in the practice (to distinguish her from someone who is able to make the moves, but only accidentally, or to distinguish a player who makes a mistake from a nonplayer). One way of making this connection, we decided, would be if the individual were able to appeal to the rules in justification, criticism, or correction of her moves.

But it makes no sense to attribute knowledge to an individual of the putative rules governing the general activities that are presupposed in rule-following in order to explain her ability to participate in these activities. This is because following a rule for any of the activities required in order to follow a rule *presupposes* the very abilities that the knowledge is supposed to explain. If someone did not know how to follow a rule, then showing her another rule would not help, since presumably if she lacked the ability to follow a rule, she would not know how to follow this other rule either. If, however, she knows how to follow a rule, then there is no point in attributing to her knowledge of another rule to explain this ability, for certainly if an explanation were required in the first place, it would be as well for the same ability that is necessary for the explanation to get underway.

In particular, if an individual does not know how to interpret symbols then showing her meaning theorems can be of no help until it is explained how any symbols in which the rules are delivered are themselves understood or until it is explained how she comes to be aware of the demands that they make.[6] If an individual does not know how to make inferences then introducing her to inference rules can be of no help until it is explained how they are to be applied.[7] If she does not know how to act for reasons, then invoking principles of rationality can be of no help until it is explained how they are to be acted upon.[8] And if she does not know how to self-ascribe mental concepts, then having knowledge of a theory of mind can be of no help until it is explained precisely how she is to use the theorems to guide her. And so forth. If, however, she already has the ability to interpret symbols, make inferences, self-ascribe, or act rationally, then there does not seem to be any point in attributing to her knowledge of putative rules governing these abilities, because her ability to follow these rules would presuppose the very abilities that are to be explained.[9]

We have a puzzle. The ability to speak and understand a language, to understand oneself and others, and to act rationally are abilities that seem to demand some sort of explanation. These abilities—which are presupposed in rule-following, insofar as they involve implementing one's understanding of a rule in action—seem to be norm-governed and yet do not seem to be governed by the same rules that govern any

6. See Sellars, "Some Reflections on Language Games."

7. A related worry about justification is raised in Lewis Carroll, "What the Tortoise Said to Achilles," *Mind* 4 (1895): 278–280.

8. This idea is developed further in part 3 and also in "De-Individualizing Norms of Rationality" (chapter 1).

9. I doubt these abilities are in any interesting way separable; I list them in order to draw attention to the scope of the claim being made. Perhaps some of the confusion about what kinds of explanations are possible arises because some theorists *do* think that some of these abilities can be presupposed in order to "explain" others (e.g., that inferential abilities can be presupposed in explaining linguistic abilities). I doubt this is coherent. For a related discussion, see Dummett's claim that a meaning theory must tell us all that is involved in speaking a language (which, he claims, is *the* rational activity par excellence) in Michael Dummett, *The Logical Basis of Metaphysics* (Cambridge, MA: Harvard University Press, 1991), and John McDowell's response in "In Defense of Modesty," (1987), *Meaning, Knowledge, and Reality* (Cambridge, MA: Harvard University Press, 2001), 87–107.

particular object game. One might thus attempt to abstract them from any particular object game, attempt to find the rules that govern them, and then use these rules as part of a cognitive explanation of the generally construed abilities. But we have just seen that if there is a distinct set of standards that govern these abilities they cannot be invoked as part of a psychological explanation of them. It makes no sense to effect the same kind of cognitive-explanatory relation here that we sought earlier between the person who is able to make the moves in the object game of baseball and the rules that govern those moves because of the regresses that threaten.

It might be useful to compare this regress—which is vicious—with another that is not especially problematic. Consider, for example, that in playing baseball one uses skills that require other abilities (e.g., to hit the ball requires the ability to stand up or to grasp the bat). This "layering" of skills is not a problem because those required are different from one another (perhaps they "bottom-out" with abilities that do not need to be learned or with those that do not require the obeying of rules). But the problem with the skills involved in rule-following is that they are the very ones that are constitutive of cognitive explanations. If it even makes sense to talk about rules that govern rule-following, following these rules would involve *the very same abilities* that the rules were introduced to govern. There is thus no possibility that knowledge of them could serve as an explanation of these abilities.

5. I HAVE GIVEN a simple argument to show that it does not make sense to attribute to an individual knowledge of rules in order to effect an internal connection between the moves she makes in following a rule, and whatever, if any, standards govern them. I have suggested that this presents a challenge to those interested in giving cognitive explanations of our broadly rational (including linguistic) abilities. Now, those hoping to offer psychological explanations of these abilities will complain that I have placed too heavy a burden on what is to count as knowledge. They will agree that knowledge of the standards that govern these abilities cannot be *explicitly* represented in the sense that they can be consciously consulted. But they will insist that it is possible to appeal to innate, implicit, or tacit knowledge in trying to give an explanation of them.[10]

Now, describing knowledge of the "rule-following rules" as *innate* would seem to solve the problem of how the abilities that are presupposed in rule-following might be learned. We might agree that they cannot be learned for reasons given above: to learn them we would have to presuppose the very same abilities that needed learning. If we give up the idea that these rules need to be learned, and accept that they are intrinsically part of our makeup, we could avoid this problem. But our problem is not merely that it is impossible to *learn* how to follow a rule by being taught these other rules. The problem is that knowledge of these other rules cannot explain the kind of abilities that are involved in rule-following. The arguments show not merely that the *acquisition* of this knowledge is problematic, but that the *utilization* of it (even if it is innate) would involve presupposing the very abilities that the knowledge was supposed to explain.

One might be tempted to avoid the problem by talking about the possibility of "implicit" or "tacit" knowledge.[11] The rules governing the kinds of abilities that are manifested in rule-following would be

---

10. Perhaps the primary motivation for supposing that the theoretical knowledge that purportedly explains the abilities is tacit or implicit is simply the recognition of the obvious fact that the knowledge is not something that the individual herself is usually able to articulate. Here, I explore whether invoking tacit or implicit knowledge can, in any case, avoid the regress threats described above.

11. The normal understanding is that implicit knowledge is available to consciousness once it has been made explicit and that tacit knowledge is not normally accessible to consciousness at all. Dennett suggests using "implicit representation" in a different sense: as information that is logically implied by something that is stored explicitly. See "Styles of Mental Representation," (1983), *The Intentional Stance* (Cambridge, MA: A Bradford Book, MIT Press, 1987). Both versions make the notion of implicit knowledge dependent on explicit knowledge. For discussions of tacit knowledge, see Gareth Evans, "Semantic Theory and Tacit Knowledge," (1981), *Collected Essays* (Oxford: Oxford University Press, 1985), 322–342. See also Martin Davies "Tacit Knowledge, and the Structure of Thought and Language," *Meaning and Interpretation,* Charles Travis, ed. (Oxford: Blackwell, 1986), 127–158; "Tacit Knowledge and Semantic Theory: Can a Five Per Cent Difference Matter?," *Mind* 96 (1987): 441–462; and "Tacit Knowledge and Sub-Doxastic States," in *Reflections on Chomsky,* Alexander George, ed. (Oxford: Blackwell, 1989), 131–152.

"cognitively grasped," yet not such that they are available for linguistic expression or conscious manipulation. Now, since the argument above is that there is nothing for knowledge of this sort to explain, it does not really matter *what* the vehicle of knowledge is supposed to be, whether or not it is accessible to consciousness, or what systemic/organizational role it plays. The problem that I am trying to call attention to does not have anything to do with the nature of the "vehicle." The problem is the explanatory poverty of putting the individual in any kind of cognitive relation to a "content" or norm that governs an ability that is itself required for rule-following. That is the quick response. Still, a more detailed look might be in order to see more precisely how the appeal to tacit knowledge, in particular, goes wrong.

Note that if there is to be any substance to the claim that knowledge of the norms plays an explanatory role, then this "explanatory role" cannot simply collapse into a description of what would constitute a correct performance. In particular, unless we have evidence that the agent subcognitively "grasps" the norm and evidence that it is in virtue of her grasping it that she acts in accordance with it, we would not be able to distinguish between her merely acting in accordance with the norms and her having the critical or justificatory abilities that tacit knowledge of the norms would ostensibly explain. The methodological difficulty of deciding which nonlinguistic performances are to count as "justifications" or "corrections" should not be underestimated.[12] But even if such a difficulty could be surmounted, it should be evident that the same regress problems would arise. This is because the abilities will involve "sub-conceptualizing" the norm, "sub-conceptualizing" the moves to which it applies, "sub-judging" its applicability, and "sub-acting" in the

---

12. If the notion is coherent, we ought to be able to say what could count as evidence that a person subcognitively conceptualizes the norm and guides her behavior in the light of it. But how would we distinguish this from the case where she misconceptualizes the norm, but acts in accordance with it nonetheless? And how would we distinguish, for example, the case where she succeeds in conceptualizing the norm yet fails to implement her understanding of it in action from the case where she simply fails to act in accordance with it? The difficulty in answering these questions puts pressure on the very coherence of the notion of "subcognitive conceptualization." For similar doubts, see Willard V. O. Quine, "Methodological Reflections on Current Linguistic Theory," *Synthese* 21 (1970): 393.

light of it. To perform these tasks *correctly* presupposes the very same abilities, operating subcognitively, that these tacit representations are, on the hypothesis we are considering, posited to explain.

Someone might be tempted to respond that, at the level of tacit representations, the question of (in)correct manifestations of abilities does not arise. She might suggest that "tacit knowledge" should be understood simply as whatever can be read off a person's dispositions under certain circumstances, where the circumstances themselves are circumscribed in such a way as to ensure that error will be ruled out. Thus, she might argue, the standards governing the abilities that are manifested in rule-following might be seen as recoverable from, or embodied in, laws of human psychology or as part of the "hardwiring" of individuals who act in accordance with them.

But this picture is confused. On the one hand, it attempts to include the role of a *norm,* whose whole *raison d'etre* is to allow us to make (logical) space for moves that are in violation of it, and yet are still made within the activity or practice that these norms in some sense govern. On the other hand, it attempts to force the opposite picture so that the existence of the norm, or knowledge of it, somehow determines that the moves will be in accord. In specifying the conditions under which the dispositions are to count as manifesting tacit knowledge, the theorist would be ruling out, for example, the cases in which someone fails to act in accordance with the norm. But this would be ruling out too much. We introduced norms in the first place, not only to distinguish individuals—as players or nonplayers—who act in accordance with the rules, but also to distinguish individuals—as players or nonplayers—who fail to act in accordance with a norm. The suggestion never was that if we could attribute knowledge of the norms to the individual we would thereby ensure that she was playing the game correctly, or even playing the game at all; it was merely that if we could attribute to her knowledge of the norms, we might have at least a prima facie case that she was playing it.

To sum up, the cognitivist faces a dilemma. On one horn of the dilemma, the subject acts in a way that can be described as "correct" or "incorrect," in which case trying to explain her ability to make correct moves by attributing to her knowledge of a putative rule governing her performances goes badly wrong when the moves involved are those within the game of rule-following. A vicious regress ensues *whether the knowledge is attributed to the person or to one of her subsystems.* On the second horn of the dilemma, the subject's performances are not apt for criticism. In this case it is not clear why we would invoke a standard in the first place, let alone attribute knowledge of it to her. This dilemma would seem to threaten attempts to provide a cognitive explanation of any of the general abilities that are constitutive of rule-following.

In fact, rule-following explanations will not even get off the ground unless we can presuppose that the individual to whom the knowledge is attributed knows what to do with this knowledge; unless we can presuppose, that is, that she has the conceptual, inferential, cognitive, linguistic, or in short, the rational abilities that following a rule requires. But the arguments above show that any attempt to give cognitive explanations of these presupposed abilities themselves is doomed to failure precisely *because* they are presupposed. These reflections ought to raise prima facie doubts about the coherence of offering cognitive explanations of the ability to understand or speak a language or the ability to understand and predict rational performances.[13] In the next section,

---

13. There are a number of prima facie targets, including:

1) those philosophers who suggest that a fruitful task in philosophy of language would be the construction of a theory of meaning knowledge of which would (suffice to) explain a person's ability to understand and speak a language. I suspect that the discussion in this chapter uncovers at least a prima facie tension in some of the requirements that Dummett, for example, places on a theory of meaning. I have in mind the constraints that a theory of meaning should "describe, without making any presuppositions, what it is that we learn when we learn to speak" (Dummett, *The Logical Basis of Metaphysics,* 9) and that the knowledge ascribed to the speakers is genuine, propositional knowledge and not a mere theoretical representation of a practical ability. These suspicions, of course, have to be examined in detail. See Dummett, *The Logical Basis of Metaphysics* and "What is a Theory of Meaning? (II)," in Holtzman and Leich, *Wittgenstein: To Follow a Rule,* (London: Routledge and Kegan Paul, 1981), 99–137, and McDowell, "In Defense of Modesty." For a discussion of the possible difficulties with theories of meaning in the light of Wittgenstein's reflections on rules, see Crispin Wright, "How Can the Theory of Meaning be a Philosophical Project?," *Mind and Language* 1 (Spring, 1986): 31–44; "Theories of Meaning and Speaker's Knowledge," (1986), *Realism, Meaning and Truth,* 2nd edn. (Oxford: Blackwell, 1993),

I shall explore in more detail how these reflections might be used to challenge the idea that principles of rationality might be invoked as part of a cognitive explanation of rational action and how this affects contemporary accounts of what is involved in acting for reasons.

III. Reason-Explanation

6. It was suggested that attributing to an individual knowledge of the rules that govern some practice might figure in an explanation of the person's ability to participate in it, but it is not at all clear *how* such an explanation would work. We wanted to seek some kind of internal connection between the rules of baseball and the person who makes the material moves so that we could say that she is *making the moves as*

204–238; and "Wittgenstein's Rule-Following Considerations and the Central Project of Theoretical Linguistics," (1989), *Rails to Infinity: Essays on Themes from Wittgenstein's Philosophical Investigations* (Cambridge, MA: Harvard University Press, 2001), 170–213;

2) linguists and psychologists, following Chomsky, who believe that it is the task of theoretical linguistics to formulate a grammar (a set of rules or principles) the tacit knowledge of which would explain a speaker's competence. See, for example, Noam Chomsky, *Reflections on Language* (London: Fontana/Collins, 1976); *Rules and Representations* (Oxford: Blackwell, 1980); see also George, *Reflections on Chomsky*, for philosophical discussions surrounding the general theme;

3) a corresponding group of theorists from philosophy, experimental and developmental psychology, and cognitive anthropology who have suggested attributing to adults a theory of mind in order to explain their ability to understand and predict rational behavior. See, for example, Martin Davies and Tony Stone, eds., *Folk Psychology—The Theory of Mind Debate* (Oxford: Blackwell, 1995); Janet W. Astington, Paul L. Harris, and David R. Olson, eds., *Developing Theories of Mind* (Cambridge: Cambridge University Press, 1989); and Andrew Whiten, ed., *Natural Theories of Mind* (Oxford: Blackwell, 1991);

4) philosophical discussions of self-knowledge that suppose a person's special access to her own mind genuinely explains her ability to self-ascribe mental states. I discuss some of the issues related to this in part 4. Boghossian and Wright both note and take on board the regress threat posed to traditional accounts of self-knowledge in Crispin Wright, "Wittgenstein's Later Philosophy of Mind: Sensation, Privacy, Intention," (1989), *Rails to Infinity*, 291–318, and Paul Boghossian, "Content and Self-knowledge," *Philosophical Topics* 17 (1989): 5–25.

part of the game of baseball. We agreed that we might be satisfied if she were able to cite the rules of baseball at the appropriate time, say, in an attempt to justify a material move. And we agreed that perhaps there would be nothing objectionable in describing this as a type of knowledge: someone who is able to cite the rules of baseball at the appropriate time might be said to *know* the rules. Then it was suggested that a person's knowledge of the rules of baseball might figure in an explanation of her ability to play it.

We could object that *if* we require that the ability to justify or correct one's moves in the light of the rules is necessary for one to be considered a real baseball player, then at best, this subability, or knowledge how to appeal to the rules to justify, etc. would be part of a complete description of the (full-fledged) ability to play the game, and not available as an explanation of it. Still, let us pursue the idea that an individual's knowledge of the rules might figure in an account of her ability to play the game. It might be supposed, for example, that knowledge of what the game requires would explain *why* the player made a certain material move. Her having this knowledge might be part of a reason-explanation for making the move.

Suppose that reason-explanation works in virtue of there being a logical or justificatory relation between a sentence describing the action and the sentences describing the beliefs and desires that are attributable to the agent.[14] We could say the individual we are considering makes a particular move (she runs to first base) because she wants to play the

14. This version of what is involved in reason-explanation is accepted by most of the people whose views I go on to criticize, though it seems to need reconsideration in light of the conclusions of this section. (I do this in part 2.) The important point is that on this model (unless it is supplemented with causation) a reason does not determine, or provide a sufficient condition for, the action that it rationalizes. Note that nothing about the ensuing argument will change significantly if values, judgments, and intentions (or statements expressing them) are added to the model of reason-explanation. Because on this more complex model, either intentions do not determine actions or all things considered judgments do not determine intentions. See David Pears, *Motivated Irrationality* (Oxford: Clarendon Press, 1984), and Donald Davidson, "How is Weakness of the Will Possible?," (1969), *Essays on Actions and Events* (Oxford: Clarendon Press, 1980), 21–42. It is this gap that reason-explanation exploits; it is this gap that makes irrationality possible.

game of baseball, and she knows that the rules require her to run to first base under certain circumstances (and she knows that these are the circumstances, etc.). This might give us at least one way of cementing the internal connection we are looking for.

Next we noticed that just as one might sometimes make mistakes in playing baseball, so, too, might one sometimes make mistakes in following its rules. This led us to suppose that one might sometimes make mistakes in following rules generally construed, abstracted from any particular object game like baseball. One might, for example, misperceive, misintuit, or misread a symbol expressing a rule, one might mistake a situation as one that falls (or does not) within the domain of the rule, and one might fail to act in accordance with the rule, even if one understood it and its domain properly. This invited us to construe the abilities that are involved in rule-following on the analogy of a "game": i.e., as activities that are themselves rule-governed.

However, even though it seems to make sense to say that someone might make a mistake in following these rules, and that avoiding this sort of error might be considered an achievement, it is difficult to see how this could be understood as a cognitive or a rational one. On the model we are presently considering, knowledge would be attributed as part of a reason-explanation for making the material moves, and the material moves we are now considering are those made within the game of rule-following. The motivation for attributing reasons here would be to effect an internal connection that would allow us to see the moves made within the rule-following game as reasonable instead of accidental. According to the hypothesis we are considering these moves would be reasonable only if the individual has reasons for playing the rule-following game.

But rule-following, under the proposal we are considering, *is* acting for reasons. Thus, to generalize the abilities constitutive of rule-following and to construe them as subject to standards would require us to invoke reasons in order to see the moves made within the *game of reason* as themselves reasonable instead of accidental. According to this proposal, an individual's moves would only be *reasonable* if she has reasons for acting reasonably. Perhaps this would require the introduction of "norms of rationality": a person might be said to have reasons for acting reasonably only if she knows what the norms of rationality require and she wants to act in accordance with them. Indeed, some have supposed that our psychological/interpretive practices are governed by norms or principles of rationality, and some have argued that these norms are contained in a theory, knowledge of which explains the ability to participate in these practices.[15]

Knowledge of these ostensible principles, however, could not figure in a substantive explanation of a person's ability to act rationally. This line of approach, like the approach in part 2, also invites a regress. The regress is vicious because the moves the individual makes that ostensibly need explaining by attributing to her knowledge of the rules of reason are exactly the same type of moves that she would be making if she were to follow them. So, again, the kind of moves that were thought to need explaining are presupposed in the very kind of explanation on offer. If it were necessary to make out an internal connection in order to obviate mere accidental rule-conformity in the explanandum, then it would be necessary to make out an internal connection in order to obviate mere accidental rule-conformity in the explanans since the moves are of exactly the same type.

We seem to have gone wrong in supposing that we need to attribute reasons for acting reasonably to a person in order to see her actions as reasonable. *If* we must effect an internal connection between reasons and action, then being able to describe a person's moves as an action (let alone in a way that displays the right kind of logical connection with statements expressing her beliefs and desires) *is* to effect it; we do not *also*

---

15. See Davidson's discussions of interpretation in *Inquiries into Truth and Interpretation* (Oxford: Clarendon Press, 1984) and the large amount of literature discussing his interpretation strategy. See also Donald Davidson, "Paradoxes of Irrationality," (1982), *Problems of Rationality* (Oxford: Clarendon Press, 2004), 169–188, in which he introduces the "principle of continence" as an example of a norm of rationality and Annette Baier's response in "Rhyme and Reason: Reflections on Davidson's Version of Having Reasons," in *Actions and Events—Perspectives on the Philosophy of Donald Davidson,* Ernest Lepore and Brian McLaughlin, eds. (Oxford: Blackwell, 1985), 116–129. For discussions involving knowledge of a theory of mind, see, for example, Martin Davies and Tony Stone, eds., *Mental Simulation* (Oxford: Blackwell, 1995), and Davies and Stone, *Folk Psychology—The Theory of Mind Debate* (Oxford: Blackwell, 1995).

need to see her actions as acquiring an independent status *as reasonable* by attributing to her knowledge that governs how to reason reasonably.[16]

7. In response, one might claim that this line of argument shows the impossibility of giving cognitive explanations for rational abilities and their kin only if it assumed that cognitive explanations are exhausted by reason-explanation, and only if it assumed that reason-explanation fits the model just described. On this model, there is a logical gap between what the agent has reason to do and what she will do. And, of course, if there is the logical latitude to fail to act in accordance with one's reasons, then attributing reasons for acting reasonably will not work, since this, too, involves a logical gap. But, so the response continues, this does not mean that we cannot explain a person's ability to act rationally on a different model. We could, for example, make scientific hypotheses concerning the laws in accordance with which she operates.

Indeed, the response might continue, if reason-explanation were to be truly explanatory, we would have to extend its domain by introducing laws or something like them into the picture anyway. We would have to do this since it has just been shown that reason-explanation has to make certain presuppositions (e.g., that a person acts rationally) *and* it has been acknowledged that a person's rational abilities might go wrong. That they do not go wrong in a particular case cries out for explanation. It cannot be a kind of explanation that itself leaves room for error or regress ensues. So rather than attempt to explain the person's rational action with higher-order rules, we might cement the connection between reason and action with causation. On certain widely accepted assumptions about the nature of causal explanation, we would thereby obtain explanation involving nomological subsumption.

Providing this sort of relation between reason and action, however, destroys the possibility of error, and it was this possibility that encouraged us to adopt the analogy of a game. Without the analogy of the game, we have lost the reason for seeking an internal connection, and

16. See "De-Individualizing Norms of Rationality" (chapter 1).

with it, we have lost any reason to attribute knowledge of the reasons.[17] To see this, consider that the possibility of psychological laws (strict or not) depends on the possibility of true, nonempty sentences expressing generalizations couched in psychological terms. Of course, rational people do not always act rationally and thus any putative psychological law will not always hold. The common response to this is to argue that irrationality is not a problem. For just as we attempt to specify natural laws by circumscribing the set of conditions under which they hold, we might do the same for psychological ones. Among the conditions to be ruled out are the circumstances in which a person is irrational. Of course, these conditions would have to be specified independently—or in a way that does not reintroduce the laws themselves—in order to ensure the truths they express are nonempty.

But even if it were possible independently to specify the conditions under which a person is, for example, self-deceived or akratic so that exceptionless laws or generalizations could be produced, ruling out cases of irrationality would be ruling out too much. For when a person acts *irrationally* as opposed to non or a-rationally, she is acting in a way that is, at least to some extent, rational. Or, returning to the analogy of a game, she is making a move in the "game" of action that can be described as recalcitrant (or "incorrect") in the light of some or other norm of rationality. It is still, however, a move within the game of action, just as an *error* is a move within the game of baseball. But if the psychological laws or generalizations are meant to apply only when someone acts rationally as opposed to irrationally, or only when someone acts *correctly* in making moves in the game of action, then a major subset of those moves we consider to be relevant will have been left out. In any case, unless error of this kind is possible, there is no reason to construe rational moves as norm-governed as opposed to law-subsumed or pattern-instantiating and thus no reason to suppose cognitive grasp of the norms is necessary to explain anything.

---

17. In "How to Resist Mental Representations" (chapter 9) I attempt to develop this argument in a way that challenges the existence of content-bearing, causally efficacious states.

The upshot is this. First, a person's acting rationally cannot be explained by attributing to her knowledge of higher-order reasons (norms of rationality) in the service of providing a reason-explanation of her reasonable action. Second, it cannot be explained by adverting to laws if they are thought to provide a determinate connection between an agent's reasons and her actions.[18] The appeal to higher-order reasons leads to a vicious regress and the appeal to laws that subsume the causal processes leads to the dissolution of the normative nature of the phenomenon to be explained. These considerations suggest that there is something wrong with the idea that reason-explanation is deficient unless it provides a way of securing a law-like connection between what the agent believes, desires, and does (or statements describing this). Reason-explanation was never supposed to provide the wherewithal to make strict predictions about what a person will do, because even if we know everything there is to know about what she wants and believes (which certainly is not required for reason-explanation) she still might act impetuously (nonrationally) or for other reasons she values less (irrationally). Reason-explanation was meant to provide us a way of understanding her actions by showing how they would make sense in the light of certain facts. But it does not (and should not) purport to tell us that she *will* act in a way that is understandable in the light of these facts.[19]

---

18. A fortiori it cannot be explained by the obtaining of causal relations between mental events, if this relation is supposed to cement the logical gap that exists between an agent's reasons and her actions. I suggest elsewhere that this motivation lies behind Davidson's claim that something is missing in reason-explanation if we consider only the purely justificatory relation between reasons and action and that this intuition motivated his introduction of a "causal element" into his account of reason-explanation. I also argue that his doctrine of anomalous monism—which commits him to the view that the only way to introduce laws into an account of mental causation is via a physical description of mental event-particulars—does not protect him from the charge that in buying into a nomological account of psychological explanation (however the laws are described) he is not leaving space for the possibility of error. See "Why Reasons May Not Be Causes" (chapter 5). Note that nothing is materially changed in the argument if the "determinate connection" is thought to hold between overriding reason and action.

19. I have not (here, or in the rest of the volume) distinguished explicitly between different kinds of standards. Dworkin makes a logical distinction between rules and principles

IV. Reconsidering Attributions of Knowledge

8. PERHAPS WE should reconsider the game of baseball to see why it was thought that attributing knowledge of the rules was necessary in the first place.[20] Again, we were seeking an internal connection between the rules and the person who makes the moves that accord with them so that we could say that she is making the moves as part of the game of baseball. This was because it was thought possible for someone to make the moves of baseball (say, run to first base) even though she knew nothing about the game and should not be credited with playing it. We thought that we might be satisfied that the performance is not accidental if the player has the ability to cite the rules of baseball in an attempt to justify or correct her moves. It was suggested that perhaps there would be nothing objectionable in describing this internal connection as a type of knowledge: someone who is able to cite the rules of baseball at the appropriate time might be said to know the rules. It was then suggested that attributing knowledge of the rules of baseball to someone might figure in part of an explanation of that person's ability to play the game. This seemed reasonable if the knowledge was to figure as part of a reason-explanation for her making a particular move. All of this seemed straightforward in the case of baseball, but we ran into trouble when we considered attributing knowledge as part of an

in his "Is Law a System of Rules?," in *The Philosophy of Law,* Ronald M. Dworkin, ed. (Oxford: Oxford University Press, 1977). Rules dictate or determine results even though they admit of exceptions; an accurate statement of the rule would take these exceptions into account. If a contrary result is reached, then the rule must have been abandoned or changed. Principles are like reasons: they incline a decision one way, though not conclusively. They do not necessitate a particular result, and they survive intact when they do not prevail. The discussion above, then, might be construed as suggesting that the norms governing our broadly rational practices are principles instead of rules.

20. It has been assumed that the rule-follower (as opposed to the mere rule-conformist) must grasp the rule, understand the situation in which it applies, and act in the light of her understanding of it. Rather than exploring in more detail what it would take to follow a rule (a central case of which occurs when one is able to cite a rule in the justification or correction of a "material" move) the discussion has centered on the circumstances in which it is necessary to invoke the distinction between rule-conformity and rule-following in order to credit an individual with making a legitimate move.

explanation of the abilities involved in the game of rule-following itself because these abilities are presupposed. But this criticism relates only to the attempt to give a cognitive explanation of the abilities that are presupposed in rule-following. Where does this leave cognitive explanations in general, and explanations involving implicit, or tacit knowledge, in particular?

Let us look at baseball one more time. In allowing that someone's ability to justify or correct her moves might warrant the attribution of knowledge, we agreed that all we meant by this is that she knows how to appeal to the rules to justify or criticize certain moves. So, *if* we decide more is needed to establish that a person is playing the game of baseball than the fact that she acts in accordance with its rules and *if* we decide that this "something more" can be supplied by attributing to her knowledge of the rules, then all we are entitled to say so far is that such knowledge (knowledge how to appeal to the rules to justify, etc.) would be part of a *description* of the full-fledged ability to play the game.

Someone might argue that such knowledge is not necessary for playing the game, if what we require is that she be able to cite the rules and if this means that she have certain linguistic abilities. It would be enough, someone might argue, to credit her with playing by the rules, even if she had no linguistic abilities. Surely we could imagine a prelinguistic child playing baseball; or, better, think of the prelinguistic children who play chess. It is very tempting to slide from this to the conclusion that prelinguistic children who play chess have implicit knowledge of the rules of the game.

But if we acknowledge the propriety of describing prelinguistic children as playing baseball or chess, this gives us a reason to retract the requirement that the child have the ability to cite the rules in justification, etc., of her moves. Since this is what we meant by "knowledge of the rules," it gives us a reason to retract the requirement that she have this knowledge. Perhaps we might still seek to assure ourselves that it was no accident that the child was playing in accordance with the rules of baseball or chess. We might rule out accidental performance if the child is able to repeat her successful performance. Or we might rule out accidental performance if the child has been trained by someone who is able to play the metagame (someone with linguistic abilities who is able to cite the rules in justification and criticism), or by someone who herself has been trained by such a person. But in imagining this, we have ruled out accidental rule-conformity without attributing knowledge of the rules to the individual who makes the moves. Again, the fact that we do not always require language-dependent justificatory abilities of someone in order to construe her as a participant in some rule-governed activity gives us a reason to retract the requirement that she know the rules—not to attribute the rules to her as the content of implicit knowledge.

Still, it might be argued that all we *mean* when we attribute implicit knowledge of the rules to someone is that she acts in accordance with the rules, and this is no accident since she was trained to act in accordance with them or because she is able to repeat her performance. So far, there is still nothing wrong with describing her as having implicit knowledge. Note, however, that training is not necessary for someone to be considered a participant in some rule-governed activity. (In order to bring this point home we will have to change the example.) An individual might be able to solve Rubik's Cube, even if she had not been trained by anybody. Indeed, someone might be able to solve the puzzle even if she could not repeat the performance. It is very tempting to slide from this to the conclusion that the person has implicit knowledge of the rules for solving Rubik's Cube, where "implicit knowledge" cannot simply refer to an ability that has come about as a result of training or to the fact that successful performances can be repeated.

But if we agree with the thought that someone might be able to solve Rubik's Cube even if she had never undergone training, then this gives us a reason to reject the idea that there must be an internal connection between the rules that govern an activity and the individual who makes the moves. We can say that it is sometimes enough to credit someone with playing the game if she acts in accordance with the rules. Knowledge (implicit or otherwise) has dropped out of the picture. To insist that someone cannot solve the puzzle unless she somehow conceives the rules (even if she cannot articulate them, even to herself) and acts in the light of her conception of the rules is simply dogmatic. What would justify such insistence? If this person were suddenly entered in a contest and produced the cube with the colors in the right places, we would not withhold the prize because she merely acted in accordance with, but did not follow, the rules. Acting in accordance with the rules *is* solving the puzzle in certain cases. Of course, the story could be filled in to make it plausible that she was not solving the puzzle at all; say she was color-blind and had the habit of fidgeting nervously with objects of any kind. This might give us a reason for requiring of *her,* or others like her, more than just the ability to act in accordance with the rules. Although it would be in order to state what additional abilities would be required in particular circumstances such as these, it would not be in order to import the requirement under the guise of "implicit knowledge of the rules" to other circumstances in which the manifestation of such additional abilities would not be required.

To insist that someone must conceive the rules *somehow*—even if what it would be for her to conceive these rules is inaccessible to us—is misguided; it fails to explain anything. Recall that one reason for supposing that "grasping the rules" was necessary was to allow us to rule out the possibility that an individual who acts in accordance with the rules does so accidentally. But if she conceives them incorrectly then she probably will not act in accordance with them. And yet if she cannot help but conceive and act in accordance with them correctly, then the possibility of mere accidental rule-conformity would be blocked from the beginning, and again there would be no point in attributing to her knowledge of the rules.

Sometimes, in certain circumstances, we might require more than the mere ability to act in accordance with the rules that govern a practice in order to rule out mere accidental conformity. We might be satisfied if the individual in question is able to play the metagame and cite the appropriate rule in the appropriate circumstances (think about exams in school). But this is not always the case. Somebody could have the ability to cite the rule in the appropriate circumstances even if she did not have the ability for which citing the rule was supposed to be evidence (suppose she cheated on an exam). Sometimes we do not need explicitly to rule out the possibility of accidental conformity, for this may be a presupposition of the particular circumstances (think about contests involving skill).

These reflections suggest that the conditions under which we credit an individual with participating in a norm-governed practice vary. When the practice is thought, action, or language the conditions under which we credit an individual with *thinking, intending, understanding,* or *meaning* vary. The criteria for applying these concepts do not remain fixed when the circumstances surrounding their application change. Problems arise when this is forgotten. For instance, if one imagines that in order to be credited with meaning something by a sign, one must have relevant justificatory abilities, then in the cases where no obvious examples of this occur in the normal way through language use, one might be tempted to describe these as cases where something like language use is going on prelinguistically nonetheless. If one imagines that to be credited with thinking, one must have the kinds of abilities that are evinced when one deliberates consciously, then one might be tempted to describe cases in which thinking seems to occur (without deliberation) as cases of unconscious or tacit deliberation. I have suggested in part 4 that in many cases involving the attribution of tacit knowledge it will be sufficient to describe the achievement as one that accords with a rule. It might not be necessary in some of these cases to suppose that the rule is internalized, conceptualized, and followed by the one whose actions conform to it. I have argued in parts 2 and 3 that for certain abilities—those that are presupposed by the kind of explanation on offer—the supposition is not even coherent. Arguments like this ought to give some support to the view that epistemic or rational norms are part of the "bedrock."

CHAPTER FOUR

Real Rules

Crispin Wright has for many years expressed frustration at Wittgenstein's "quietism"—his refusal to offer substantive answers to the metaphysical and epistemological problems that are raised, Wright alleges, by Wittgenstein's own reflections on rules. In his most recent paper, Wright suggests this quietism can be explained by Wittgenstein's rejection of a picture that seems to indicate Platonism and communitarianism as the only available solutions to these ostensible metaphysical and epistemological problems.[1] I agree with Wright that Wittgenstein would reject the initial assumptions that pit the realist against the communitarian, but I tell my own story on behalf of Wittgenstein about what is wrong with the altogether misconceived picture that generates the dilemma.[2]

---

1. Crispin Wright, "Rule-Following Without Reasons: Wittgenstein's Quietism and the Constitutive Question," in *Wittgenstein and Reason,* John Preston, ed. (Oxford: Blackwell, 2008). See also Wright's *Rails to Infinity-Essays on Themes from Wittgenstein's* Philosophical Investigations (London and Cambridge, MA: Harvard University Press, 2001).
2. This chapter develops ideas from "Playing the Rule-Following Game" (chapter 3) and "Une Cartographie des Concepts Mentaux," Critical Introduction to Gilbert Ryle's *La Notion d'Esprit* (Paris: Payot, 2005), 7–70, reprinted as "Rethinking Ryle: A Critical

1. In §31 of the *Investigations,* Wittgenstein makes a few remarks about rules when he asks what one has to know in order to be capable of asking a thing's name. He suggests, among other possibilities, that a person may have learned the game, and become a master of it, "without learning or formulating rules." In §54 Wittgenstein asks us to recall the kind of case in which a game is played according to a definite rule. A rule, he suggests, may be an aid in teaching the game or it may be an instrument of the game itself.

> Or the rule is employed neither in the teaching nor in the game itself; nor is it set down in a list of rules. One learns the game by watching how others play. But we say that it is played according to such-and-such rules because an observer can read these rules off from the practice of the game—like a natural law governing the play.[3]

It seems, then, that we may say of a game which its participants have mastered that it is played according to such-and-such rules because an observer can read the rules off from its practice. Picking up the remark in §31, we can add that the participants who have mastered the game may know its rules other than by having learned or formulated them: borrowing an expression from Ryle, we can say, in such a case, that they know the rules "by wont."

Here is a mythical picture that needs resisting. What is used in the teaching of games, consulted in the course of the game, or read off by an observer, are mere expressions of rules: the real rules are something at which these expressions only gesture. Once grasped, apprehended, or

---

Introduction to *The Concept of Mind,*" in Gilbert Ryle, *The Concept of Mind* (London: Routledge, 2009), ix–lvii. For a different but complementary discussion (as well as an interest in chocolate and eggs) see David H. Finkelstein "Wittgenstein on Rules and Platonism," in *The New Wittgenstein,* Alice Crary and Rupert Read, eds. (London: Routledge, 2000), 53–73.

3. To the question how the observer distinguishes mistakes from correct play, Wittgenstein suggests that there are characteristic signs of having made a mistake (or making a correct move) in the players' behavior. This is important, because it reminds us that what is "read off from" the practice of the game is not every move a player makes, but only those which help set the standard for what counts as correct and therefore as incorrect.

intuited by a participant of the practice there is no rational option but to do as the rule requires: apprehension of the rule is sufficient to determine and thus to explain how the one who grasps it acts as it mandates. According to this mythical picture, when expressions of the rules are not used in the learning or teaching of the game, nor within the practice itself, those who have mastered the practice can be credited with more than knowledge by wont of its rules: they have (somehow) come into cognitive contact with the rules, unexpressed or unrepresented though they may be, which (somehow) guide them and thus (somehow) explain their ability to act as they mandate. A "cavalier" realistic version of this myth supposes that the real rule, shorn of its expression, exists independently not only of any particular individual's grasp and propensity to act in accordance with it, but independently of human practices altogether: the rule consists in steps that are already drawn well in advance. A noncavalier version would, presumably, accept that rules are generated from human practices but insist that they take on a life and an independence of their own: here too, the steps are still in a sense drawn in advance.

My diagnosis (on behalf of Wittgenstein) of what inclines us to believe either version of this mythical picture is that we tend to forget that rules are implicated in our practices in a variety of ways; that, correspondingly, there are different things we mean by the expressions "knows the rules" or "obeys the rules"; and in particular, we forget that one of the ways rules are implicated is *merely* that an observer can read them off from the practice in question. Forgetting this, we choose for our paradigm the occasions in which it makes sense to say the agent is guided by her apprehension of a rule-representation. When confronted by the fact that in a large number of cases, the participants consult no sample, table, definition, signpost, formula, etc., we (mistakenly) suppose that rule-guidance occurs via the immediate apprehension of an unrepresented rule. Forgetting that the activity of following a rule-representation comes replete with a catalogue of its own kind of errors, we accommodate the absence of errors by (mistakenly) supposing that this unrepresented rule guides us to action via perceptual and rational mechanisms that are guaranteed to succeed.

This mythical picture, which continues to dazzle philosophers (and which Wright still struggles to resist), does indeed present deep metaphysical and epistemological puzzles. But it results from a conflation of two different senses of "knowing the rules." It conflates "knowledge by wont" of the rules, which an observer can read off from the practice of the game, with the cases in which rule-representations do genuinely, but all too fallibly, guide the participants. Many of the alleged philosophical mysteries about the constitution and epistemology of rule-following begin to unravel when we take seriously these different senses. In particular, puzzles about the nature or constitution of rules, their creation, their objectivity, and their relation to the assent of a community evaporate when we recognize that rules may be implicated in our practices merely in the sense that an observer can read them off from the practice in question. Questions about how they lead us at all, let alone in an open-ended range of situations, are answered by taking a look at how they are expressed and how they are appealed to—on those occasions when they are—in the teaching or in the course of the practice.

2. MANY OF THE activities in which we engage consist in performances or moves which are sometimes required, sometimes prohibited, and sometimes merely allowed or permitted, depending on the activity it is and the circumstances in which it takes place. References to the performance rules that govern these activities are "references to criteria according to which performances are characterised as legitimate or illegitimate, correct or incorrect, suitable or unsuitable, etc."[4] Sometimes the rules may guide us (when expressions of the rules are appealed to in the course of the game); sometimes there are no expressed rules to do any guiding: in such a case, rules are implicated in the sense that they can be read off from the practice like a natural law governing the play. In the latter case, we allow that they are performance rules but only

---

4. Gilbert Ryle "Why Are the Calculuses of Logic and Arithmetic Applicable to Reality?," (1946), *Collected Papers,* vol. 2. (London: Routledge, 2009), 238.

because they are the criteria by which moves are judged to be correct or incorrect, etc.—not because they have been consulted in the course of play. The criteria thus crystallized from successful performances are those of an observer: a theorist, say, or a referee.

Crystallizing the moves that work and distinguishing them from those that do not provides the theorist a short-hand way of showing that the move was appropriate, required, correct, or successful, etc. To adopt the stance of the theorist allows one to justify the performance or move: in showing that it accords with a rule, norm, or standard its success is exhibited; its failure is diagnosed by showing that it violates a rule, norm, or standard. Its success or failure is due not to the mere fact that it conforms to or flouts a rule or norm: the rules or norms were themselves distillations of moves that were already deemed successful.

Although there are variations upon, and exceptions to, this simple picture, it will help to imagine a somewhat artificial case in which a practice is already up and running before a theorist-observer comes along and distills the moves that work from those that are irrelevant or unsuccessful; in the course of doing so she may also invent a procedure to help initiates of the practice master the steps that she deems relevant and successful. It was assumed, for example, that enough Americans would appreciate brownies baked as described on the box of Baker's Unsweetened Chocolate (and buy the chocolate), so a recipe for "Baker's One-Bowl Brownies" was printed on the box. The brownies were deemed good enough to establish a general procedure for making them. (Similar stories could be told as to why one would put up signposts, construct maps, invent formulae, etc.)

I am ignoring, for this discussion, the cases in which practices evolve and in which the theorists (recipe-writers) are also and at the same time participants in the very practices about which they are theorizing. But using this overly simple model, we already see something right and something wrong in Wright's suggestions that 1) in any area of human activity where there is a difference between correct and incorrect practice, what counts as correct or incorrect is partly determined by rules; 2) that these rules somehow guide us; and 3) that they issue their determinate mandates or generate facts about which a response is required in heretofore unconsidered types of situations.[5]

First, it is true that what counts as success may be partly encapsulated by the rules: showing that a move does not conform to a norm or rule (or that it violates one) may be a way of signaling why it is in error. If the brownies do not turn out as planned when someone fails to add the eggs, pointing to the instruction in the recipe that says "add 3 eggs" demonstrates why. But what determines or constitutes what is correct or incorrect in this particular activity depends on how the finished product tastes, etc. and this will depend on much more than could ever be encapsulated in the recipes. The rules considered here—recipes—are merely expedients to help replicate the dish.

Second, if, as I am recommending, the only rules that are candidates for genuine consultation by participants in the practice (and not merely read off from it by a theorist) are representations such as recipes, signposts, maps, formulae, tables, samples, etc. then these representations may or may not guide those participating in the activity. Someone who adds the eggs to the melted chocolate and butter without consulting the recipe is not making the kind of mistake that would prevent him from cooking "Baker's One-Bowl Brownies" to perfection. The brownies are good in part, we may agree in our role as theorists, because three eggs are added: not (on this occasion) because a recipe is consulted beforehand.[6] So, one may point to a recipe in order to give a summary justification of a culinary move (the addition of three eggs) but this does not require that the one given credit for the performance must consult the recipe.

Third, the rules may *not* enjoin determinate mandates, permissions, and prohibitions in previously unconsidered types of situation. As

5. See Wright, *Rails to Infinity* and "Rule-Following Without Reasons."

6. During cooking lessons, as in other lessons, the student may be required to show her work or to show why a particular move was justified as she is making it. In these cases, proof of success is to be found not just in the pudding (as it were) but also in the ability to show that the pudding accorded with the recipe. But these moves are part of a "learning-to-cook game," with different conditions of success from the lower-order "cooking game."

Wittgenstein reminds us (§80), if the situation changes drastically, we may not know what to say. The recipe-instruction "add 3 eggs" on the back of the box of Baker's Chocolate does not tell us what to do if all the eggs have double-yolks or if only ostrich eggs are available.

Nonetheless, it is in the nature of rules that they may be satisfied by any number of performances which may differ in any number of ways and still count as applications of the rule. One can add the eggs to the mixture after the fashion of James Bond's dry martini by shaking and not stirring; or after John Steed's sugary tea, by stirring counterclockwise; or by adding the 100034th, 100035th, and 100036th eggs one has handled in one's lifetime (cf. §186). The game of cooking, like the application of a word, is not everywhere bounded by rules. The idea that rules generate facts seems a mystery only if we imagine that the rule-codifications come first and the practices which the rules govern are somehow generated from the rules themselves. But this is part of the mythical picture that needs resisting.

That rules may be satisfied by any number of performances is no more mysterious than the fact that it is in the nature of the formula "If *P* then *Q;* given not-*Q,* so not-*P*" to be satisfied by an indefinite number of arguments, or of the statement "All men are mortal" to be satisfied by an indefinite number of humans. We formulate rules with the aid of universal generalizations such as "whenever" or "every time"; or, we may express the rule or instruction as an open or partly open hypothetical statement that contains at least one variable. (This will make it clear that the statement is to be applied to or satisfied by various situations and that it does not describe them.[7]) It seems, then, that if we keep in mind how rules are expressed the mystery fades as to how we create rules which carry instructions for an open range of situations (to the extent they do so) that we do not think about in creating them.

Nor does it seem a mystery how and when it can have been settled that it is one specific rule in particular we are following when everything we may so far have said, explicitly thought, or done would be consistent with its being any of an indefinite number of potentially

7. See Ryle, "Why Are the Calculuses of Logic and Arithmetic Applicable to Reality?"

extensionally divergent rules.[8] When the rule is part of the teaching of the game, or consulted during the game itself, the rule is the one the teacher expresses or the participants look up. How and when it can have been settled that it is one specific rule we are following is only mysterious when we cling to the mythical picture that in participating in a practice we have somehow grasped unrepresented rules. In the case in which rules are merely there to be read off from the activity in question and there is no rule-representation consulted or used in teaching then there is *not* only one set of rules to be read off from the practice. The recipe-writer, like the mapmaker, will codify the moves she wants to teach others how to replicate by assuming a number of background abilities and a shared response.

For example, in writing the recipes, the recipe-writer ignores, adjusts, and modifies a great deal of what the chef does in order to make the procedure as fail-proof as practically possible. If she is writing for an audience whose skills are diverse she will have to make the recipes more general than she would if she were writing for experts. But she will still have to presuppose a certain amount of knowledge or her recipes would never get written. The rule, "add 3 eggs" was written for someone who already appreciates the importance for brownies of the yolk and the white and not the color of the eggshell. It was written for someone who already understands that the eggs should be discarded if there is even a partially formed chicken fetus inside, if the eggshell is cracked, if the egg is over three weeks old or if it floats in water, or if the contents have been either been sucked or blown out. And so on indefinitely.

3. Wittgenstein explores further the theme of rules in §201, but in this passage we seem to be leaving aside, temporarily, the sense of "knowing the rules" in which someone is given credit for having carried something off by a theorist who reads the rules off from the practice and concentrating instead on cases in which expressed rules (say, a signpost: cf. §198) are consulted in the course or in anticipation of the performances that constitute the practice. In such cases,

8. See Wright, *Rails to Infinity* and "Rule-Following Without Reasons."

Wittgenstein suggests, there is a way of grasping a rule which is not an interpretation, but which is exhibited in what we call "obeying a rule" and "going against it" in actual cases. This thread of reasoning is followed a few lines later by the remark: "And hence also 'obeying a rule' is a practice."

When someone comes across a signpost pointing to the right, obeying the signpost may well consist in taking the fork to the right. When someone comes across a recipe for brownies that says "add 3 eggs," obeying the recipe may well consist, on that occasion, of cracking open three eggs, and then adding the contents to the melted butter and chocolate. Producing paint in the shade of "Uzès Green" may require consulting a color chart and following the formula for adding *x* drops of buttercup yellow and *y* drops of cobalt blue to the bucket of white paint. How does the mixer of paint know how to add the drops? We assume he acts as described; we imagine he has been trained to do so. We do not need, of course, to assume that he follows another rule telling him how.

Only sometimes will obeying a rule involve something more complex: the consultation of the rule (-representation), and then puzzling over what it says for a while before acting in accord (or not). Puzzling over what the rule says may involve reflecting on various possible interpretations; this in turn may involve appealing to something (e.g., a schema for interpreting the rule.)[9] This higher-order procedure may then result (or not) in action that is in accord.

According to the mythical picture, mere grasp of the rule ensures that one acts as it mandates; this is alleged to explain one's ability to master the practice in question. But it is clear that one's apprehension of a rule-representation carries no such guarantee and thus fails to offer the kind of explanation coveted by the myth. The concept of following a rule (-representation) is a polymorphous concept in its own right, an umbrella term for a multivaried set of practices, each with moves that are sometimes required, sometimes prohibited, sometimes merely

9. See Ludwig Wittgenstein, *Philosophical Investigations,* G. E. M. Anscombe, trans. (Oxford: Blackwell, 1953), §86.

allowed or permitted. An observer can read off from the practices of following a rule (-representation) performance rules that govern this higher-order activity; a crystallization of the moves that work from those that do not may provide her with a method for justifying the moves. A distillation of the moves she deems relevant and successful may figure in a procedure that would help initiates of the various practices—those practices that require the participants to be guided by a rule-representation—to be guided successfully. Such a procedure may involve higher-order rule-representations (e.g., schemata).

But of course *these* higher-order practices, and what counts as following a rule-representation successfully within them, begin and end with knowledge by wont of following the rule-representations. For the introduction of higher-order rules, like the introduction of first-order rule-representations, will never explain the ability to master the practice in the way demanded by the myth—one which leaves no gap between the rule and its successful uptake.

One corrective to the idea that the rule itself, once apprehended, shows the way such as to guarantee correct performance is to remind oneself that represented rules present no such guarantee. The instruction "add 3 eggs" does not in any mysterious way *determine* what one should do. As we have seen, appealing to the instruction is not even necessary, for someone can cook a dish without having followed this instruction. Nor is it sufficient since this instruction, even if read correctly, can be misinterpreted or misapplied: adding the entire eggs with shells would be one such way. But adding information on how to interpret the recipe is only useful to ward against one particular misunderstanding: one, say, in which the eggs are added whole. But that does not mean that all other misunderstandings have been prevented: one could misunderstand "is correctly obeyed by cracking open three eggs and adding them to the mixture" by taking "them" to refer to the egg-shells (now cracked open) as well as the contents of the eggs. (Change the circumstances slightly so that the instruction "add 3 eggs" is given to the cook as she is setting up her ingredients and the requirement to crack them open first would be disastrously out of place.) It seems the addition of higher-order rules merely iterates the problem. There is no way of specifying how the rule is to be understood such that misunderstanding or misapplication is ruled out. Rules, insofar as they are genuinely consulted, and not merely read off from the practice by an observer, are always expressed or represented in some way, and the fact that they are represented leaves open the possibility that they are misrepresented (misread, misinterpreted, and misapplied). This goes for higher-order rules as well.

But having accepted these facts, it would be a strange reaction to say: "The rules do not therefore determine what I am to do thus I am on shaky ground and there is no such thing as a correct way of obeying a rule." Those who are familiar with the practice of cooking can tell and can teach what counts as a correct way to add three eggs and what does not. But having this ability requires that one has, to some extent, mastered the practice (or "art") of cooking. Rules are abstractions from practices up and running; obeying a rule requires that there be steps that count as instances of rule-obeyance. This is why obeying a rule is a practice. But this, of course, does not amount to a community of observers that nods its assent or registers its dissent.

Beyond the obvious everyday contingencies that we simply recognize as being similar (and for which there is no explanation beyond shared propensities and a shared form of life), we may not know how to apply a rule and the rule itself does not tell us. Whether or not to count one-eighth of an ostrich egg as equivalent to one chicken egg (for the purpose of making One-Bowl Brownies) is not implicit in the instruction "add 3 eggs." In this sense, a decision (of sorts) is required as to whether a fraction of an ostrich egg is a reasonable enough substitute. This will depend on the taste and the texture of the brownies, whether ostrich eggs are poisonous, and so forth and so on. It is not so much the interpretation of the rules that matters and about which we must make a decision, as the fact that we must take our turn as theorists and return to the practice of brownie-baking, to reconstruct the moves that work from those that do not, given the new (chicken egg-depleted) circumstances.

I agree with Wright's recent conclusion that Wittgenstein's quietism does not result from (as Wright once suggested) an uncharacteristic "philosophical turning of the back" but rather from the fact that Wittgenstein would reject the initial assumptions that set up the dispute between the Platonist and the communitarian. I have argued that different senses of "knowing the rules" tend to be conflated; that one may know the rules in the sense that one acts in accordance with rules distilled by a theorist; and that genuine consultation of rules is neither necessary nor sufficient to guarantee action that accords with their mandates. I thus hope to have contributed to an understanding of why we should reject the mythical picture that sets up the dispute: it is one that indeed misconstrues the nature of real rules.

PART TWO

Reason-Explanation and Mental Causation

CHAPTER FIVE

Why Reasons May Not Be Causes

Introduction

1. Davidson's "Actions, Reasons, and Causes" defends the ostensibly commonsensical view that rationalization is a species of causal explanation.[1] The arguments are generally considered to have put to rest Wittgensteinian anxieties about an illicit conflation of explanatory paradigms implicit in the very notion of "mental causation." In quieting these anxieties, the arguments have served as an imprimatur for subsequent generations of realists about the mental. Mental realists construe mental predicates as picking out determinate, interpretation-independent states of affairs. Realists are encouraged by these arguments because if mental causation exists, an arguably necessary condition for realism has been met—at least, that is, if Jaegwon Kim's intuition is correct that "to be a mental realist, your mental properties [say] must be causal properties—properties in virtue of which an event enters into causal relations it would otherwise not have

1. Donald Davidson, "Actions, Reasons, and Causes," (1963), in *Essays on Actions and Events* (Oxford: Clarendon Press, 1980).

entered into."[2] In particular, the thesis that reasons are causes has been embraced by realists who are physicalists about the mental. Physicalists construe mental predicates as picking out particulars or properties that can be (synthetically or theoretically) identified with particulars or properties referred to by physical predicates. Physicalists, in assuming the causal closure of physics—that we ought to be able to trace the causal ancestry of an event without going outside the physical domain, would consider the existence of mental causation to ensure the truth of their view. Finally, a conviction that some sort of physicalism will be vindicated has, in turn, provided a certain impetus to eschew a "conceptual" investigation of the mental—or an ontologically neutral investigation of the practices in which our mental concepts figure—in favor of a more "scientific" or "empirical" approach which has proliferated, mainly through the cognitive sciences, in the last thirty years.[3]

The line of thought described above is not the route traversed by Davidson, who rejects the thought that psychology is or can ever be like the physical sciences and whose own brand of (token) physicalism depends on a particular view about the individuation of events and an analysis of singular causal statements.[4] Interestingly, however, while the rest of Davidson's corpus is controversial, the original argu-

---

2. Jaegwon Kim, "The Myth of Nonreductive Materialism," (1989), in *Supervenience and Mind : Selected Philosophical Essays* (Cambridge: Cambridge University Press, 1993), 279.

3. See especially Jerry Fodor, *The Language of Thought* (New York: Crowell, 1975); *Representations* (Cambridge, MA: MIT Press, 1981) and *Psychosemantics—The Problem of Meaning in Philosophy of Mind* (Cambridge, MA: MIT Press, 1987). Indeed, convinced of the existence of mental causation, contemporary philosophy of mind presupposes that propositional attitudes are at bottom physical, and the central concern has been to specify how physical states can be bearers of explanatorily (causally) relevant content. Various naturalist reduction strategies for mental content have been proposed. See, for example, Fodor *Psychosemantics* and *A Theory of Content* (Cambridge, MA: MIT Press, 1990); Fred Dretske, *Knowledge and the Flow of Information* (Cambridge, MA: MIT Press, 1981); Ruth Millikan, *Language, Thought, and Other Biological Categories* (Cambridge, MA: MIT Press, 1984) and *White Queen Psychology and Other Essays for Alice* (Cambridge, MA: MIT Press, 1993).

4. See Donald Davidson, "Psychology as Philosophy," (1974), in *Essays on Actions and Events*.

ments supporting the thesis that reasons are causes have been widely accepted; indeed, the sweeping acceptance of the thesis has recently been described as one of the few achievements of contemporary analytic philosophy.[5] But one cannot get to any type of physicalism about the mental—nor to the view that mental predicates pick out phenomena that are apt for empirical investigation—from Davidson's argument that reasons are causes, without presupposing a paradigm of explanation that begs the question against the opponents (as, indeed, Davidson's views on actions, events, and singular causal statements do). Indeed, any inclination that one has toward the view that causation is a relation that implies this explanatory paradigm gives one a positive reason to reject the thesis that reasons are causes. Or so I shall argue.

In section 2, I shall consider Davidson's positive arguments for introducing causation into his analysis of rationalizing explanation, and leaving the notion of causation largely unanalyzed, I shall argue that the arguments do not obviously support the intuition that something is missing from a (merely) rationalizing explanation that can be provided by the thesis that reasons are causes.[6] They certainly do not entail the (robust) existence of mental processes. In section 3, I shall say what presuppositions about explanation seem to motivate Davidson to introduce causation into the analysis, and I shall argue that his views regarding the anomalism of psychology give him a better reason for rejecting his causal thesis than they give him for espousing anomalous monism. In section 4 I shall say more about why there cannot be psychological laws. I shall conclude by suggesting why reasons should not be construed as causes.

---

5. See Millikan, *White Queen Psychology.*

6. A "rationalizing" relation in Davidson is ambiguous between an explanatory relation between reasons and action that evinces a purely logical or justificatory relation alone and an explanatory relation between reasons and action that has causation already built into it. Davidson argues that rationalizing explanation is a species of causal explanation; but in order to argue for the necessity of introducing causation into the analysis, we need a way of talking about the relation before causation has been introduced and show that it is wanting in explanatory power. I shall defer to common (philosophical) usage and refer to this non-causal relation as a (mere) rationalizing or justificatory relation.

The Search for the Mysterious Connection

2. Davidson's positive argument for supposing that rationalizing explanation is a species of causal explanation, in the sense that "a primary reason for an action is its cause" rests on two claims. First, he suggests that the justificatory pattern exhibited in reasons' relation to action is not sufficient for explanation, since:

> . . . [if justification requires only] that the agent have certain beliefs and attitudes in the light of which the action is reasonable . . . then something essential has certainly been left out, for a person can have a reason for an action, and perform the action, and yet this reason not be the reason why he did it. Central to the relation between a reason and an action it explains is the idea that the agent performed the action because he had the reason.[7]

Second, he argues that this "because" relation should be construed as causal:

> If . . . causal explanations are "wholly irrelevant to the understanding we seek" of human action then we are without an analysis of the "because" in "He did it because" where we go on to name a reason. . . . But I would urge that, failing a satisfactory alternative, the best argument for [introducing a causal factor into the analysis] . . . is that it alone promises to give an account of the "mysterious connection" between reasons and actions.[8]

I would like to consider these claims in some detail.

According to Davidson, a primary reason is attributed with the aim of explaining why an individual performs an action under a certain description. A primary reason consists in a pro-attitude the agent has toward a certain state of affairs, together with a belief that the action

7. Davidson, "Actions, Reasons, and Causes," 9.
8. Ibid., 11.

(under that description) is, or is likely to bring about, that state of affairs. Davidson says,

> Thus there is a certain irreducible—though somewhat anaemic—sense in which every rationalization justifies: from the agent's point of view there was, when he acted, something to be said for the action.[9]

What more is required for rationalizing explanation than is provided by primary reasons alone? There are many possible answers to this question, which I shall examine in what follows. In considering the possible answers, I shall be asking whether or not (or to what extent) what is missing from explanation can be provided by construing reasons as causes. In a later article, Davidson gives a typical argument for introducing causation into a primary reason account:

> A desire and a belief of the right sort may explain an action but not necessarily. A man might have good reasons for killing his father, and he might do it, and yet the reasons not be his reasons in doing it (think of Oedipus).[10]

So, let us think of Oedipus as Davidson suggests.

External Reasons

Having heard from a prophet that he was destined to murder his father and marry his mother, Oedipus attempted to escape his fate by leaving the village in which he lived. As he was walking along, he was accosted by a number of men who were traveling by carriage. They got into a brawl, and Oedipus killed them; in particular, he killed an old man who hit him over the head with his staff. This man was Laius, the King, Oedipus's father. What reason did Oedipus have to kill this man? His motive was self-defense, and we can easily translate this into

9. Ibid., 9.

10. Davidson, "Psychology as Philosophy," (1974), *Essays on Actions and Events,* 232.

a rationalizing reason: he believed the old man was about to kill him, and he believed that unless he killed the old man first he would die, and he wanted to live. What is the good reason Davidson refers to that Oedipus "has" but which is not "his in acting"? Perhaps it is that Oedipus's father, who also heard the prophecy, ordered his son's murder when Oedipus was an infant, and Oedipus might want revenge; perhaps it is that if Laius is out of the way, Oedipus will be able to take his father's place in his mother's bed. In any case, Oedipus does not know that the old man in the carriage is Laius the King, or his father, or the husband of Jocasta, or the husband of his mother. So neither the vengeful motives nor the "Oedipal" motives could be construed as comprising one of the reasons for which Oedipus acted. But this does not show the insufficiency of primary reasons, since neither of these are primary reasons as Davidson defines them. That is, neither of these motives could be construed as constituents of reasons that could be attributed to Oedipus to explain his action: whether or not he has the vengeful or the Oedipal desires, he has not got the requisite beliefs to combine with these desires to rationalize the action of killing of his father. He merely has reasons which rationalize killing the old man in the carriage.

So far, thinking of Oedipus does not really help. But maybe we can try to reconstruct the example. What we need is a sense of "reason" such that a reason is properly attributable to the agent and yet still not be the reason for which the agent acts.

Overridden Reasons

So let us change the story a little. Let us suppose that Oedipus does know that the man in the carriage is his father and let us suppose that he both wants to kill his father because he ordered his murder, and he wants to kill him so that he can marry his mother, and he wants to preserve his own life from harm. Of course there ought to be nothing wrong with saying he acted for all of the reasons. If someone tells you I bought my house because it was affordable, it has a big garden, and a beautiful view it would be strange to demand the reason for which I acted. In certain cases, the more reasons we give, the more explicable the event—not the other way around. But it is easy enough to complicate the story to make sense of the idea that Oedipus has all the above reasons for acting, yet only acts because of one of them. Suppose that although Oedipus wants revenge, and would like to take his father's place in his mother's bed, he nonetheless has strong moral inhibitions against killing a parent for any reason, let alone to satisfy vengeful or incestuous desires—inhibitions strong enough to resist the temptation to kill his father for either of these reasons. But suppose, too, that Oedipus has an even stronger desire to live, and in the end, this outweighs his moral inhibition against killing his father. So he kills him. In this sense, perhaps we could argue that it would be inappropriate to cite the vengeful and incestuous reasons to explain Oedipus's action. So, perhaps in this case, though Oedipus has good reasons for killing his father, they are not the ones for which Oedipus acts.

Here is an example that shows that citing a primary reason is not sufficient to explain an action. But the more that we might add is precisely what I did add in describing the story: a more complex justificatory machinery which would allow us to attribute not only beliefs and desires (or primary reasons), but competing primary reasons, values, and weighted judgments as well.[11] But nothing yet has been argued about the necessity of positing a causal relation between reasons and actions: we just need to introduce judgments, weights, and values into the "anaemic" analysis of reasons. But these judgments need not necessitate talk of mental processes; they may simply be, as reasons and intentions are for Davidson, part of a more complete analysis of the concept of acting for reasons.

Citing a primary reason might not be sufficient to explain an action even when it is not overridden by competing value judgments. It may simply be that one primary reason "carries more weight" than the others. It may be true, for example, that my house has a big garden and

---

11. This suggestion is made in Davidson, "How is Weakness of the Will Possible?," (1970), in *Essays on Actions and Events* and in subsequent articles. But by then the thesis that reasons are causes was already in place, so that the addition of principles, weighted judgments, etc., was added to a reasons-as-causes model and not substituted for it.

a beautiful view and it was within my price range, but the reason I bought it was the garden; I should not have bought it otherwise. How are we to understand this unless we speak of this reason as being "causally efficacious"?[12] But we can accommodate this example by allowing that certain reasons, values, or desires have more weight than others. But again, to assign weights to desires, etc., is merely, according to the story I shall tell, to complicate the justificatory machinery. It does not imply the existence of mental processes.

Avowals

Perhaps we might be curious to know what the agent himself would say about his reasons for acting. This might be a different sense in which the reasons are really his. But again, if we are attributing reasons as we ought to be—so that they manifest the agent's beliefs about the situation, and his desires, etc.—then what the agent says is presumably being taken into account all along. That is, the utterances that Oedipus is disposed to make will count as (defeasible) justification for attributing or not attributing to him various beliefs and desires: that Laius is his father, that Jocasta is his mother, that he avenge himself, and so on.

Suppose that we have justification for attributing attitudes to Oedipus consistently with the story as Sophocles tells it. When he kills Laius, we will describe the action as the killing of the old man in the carriage and the motive as self-defense. Now, when we ask Oedipus, after the fact, why he killed the old man, we would expect him to say "because he was going to kill me." Suppose, though, he does not: suppose he says that his reason was that he wanted to avenge himself against a past injustice. Then, we would be entitled to ask him more about what he knows about the past injustice, what he knows about the old man, and how he knows it. Unless Oedipus could satisfy us that we had misinterpreted the justification conditions before—that is, unless he could show us that his own self-attributions were justified, and that we had made a mistake in refraining from attributing to him these beliefs—we would not have any reason to revise our interpretation. Instead, we would have

12. This was suggested by Jim Hopkins.

a reason to convince Oedipus that he was talking nonsense. So although it is true that what Oedipus is inclined to say is relevant, it certainly does not decide the matter. In any case, nothing has been established by this example about the insufficiency of primary reasons. If we have attributed them correctly, then what Oedipus is inclined to say should have been taken into consideration all along, and if it was not then it is possible we have misascribed the primary reasons. The same considerations will be relevant for attributing values, weighted judgments, and intentions. It is entirely likely that accepting some of the subject's own testimony about her reasons for acting is necessary for complex attributions. But so far this does not imply talk of mental processes.

Buridan's Ass

It might be responded that there is still a lacuna in explanation left by pure rationalizations.[13] Suppose, as suggested, our primary reason account is supplemented by attributions of values and weighted judgments. Now suppose Oedipus finds himself in a "Buridan's Ass" type situation: he finds himself at the crossroads after the brawl and has to choose which way to go. Suppose that each of the three paths leads to the same destination and requires traveling the same distance. Suppose that Oedipus knows this and though he has reasons for choosing any one of the paths (he wants to get to where it leads) he has no reason for choosing one path over another; that is, no amount of additional deliberation will deliver a stronger judgment in favor of his taking one path over the other. Now suppose Oedipus takes the path on the right. It might be thought that his having reasons for taking the right path does not provide an explanation of his action since his reasons for taking one of the other paths are just as strong. If we want an explanation of his taking the right path instead of one of the others, then (it might be argued) rationalizing explanation needs to be supplemented with a causal explanation. Oedipus's taking the right path instead of one of the others would be explicable if his reasons for taking the right path caused his action.

---

13. This objection was suggested by Christopher Peacocke.

Surely the moral of the tale of Buridan's ass is that if the ass were a rationalist, and he only acted on his preferences, then he would starve to death: or if Oedipus only acted on his preferences then he would never get to his destination. So much the worse for a rationalism that insists on the contrary. The moral is that rationalizing explanations have a limit and that when faced with a choice for which there are equally strong reasons, an individual is eventually forced to pick, not choose.[14] That is, although Oedipus has reasons for taking the right path his taking the right path instead of another cannot be rationalized.

We may well have significant results from psychological experiments showing that right-handed people tend to take options on the right. This statistical regularity is perhaps best explained neurophysiologically, so it would be natural to suppose that this story (which is intuitively causal) might be added to a rationalizing one to explain Oedipus's taking the road on the right instead of one of the others. But now we must be careful to state exactly what is being explained. For the example to be used in support of Davidson's thesis that reasons are causes, it would have to be the case that causal explanation is incorporated into rationalizing explanation. This would be achieved only if the action to be explained is Oedipus's choosing the road on the right over the others. But although the possibility of such neurophysiological explanations is quite plausible, that this is their explanandum is not. What is plausible is that the neurophysiological hypothesis explains Oedipus's picking the road on the right instead of one of the others, not his choosing it. But if he picks it (and the neurophysiological hypothesis best explains why) this is by and large irrelevant to his having reasons. Even if these statistical regularities kick in only when a person has reasons in acting (neither one of which is overriding), this does not show that providing a hypothesis that is sufficient for predictive purposes renders certain

14. By "pick" I mean take a course of option more or less arbitrarily, and by "choose" I mean take a course of option that reflects one's preferences. See Edna Ullmann-Margalit and Sidney Morgenbesser, "Picking and Choosing," *Social Research* 44, 4 (1977): 757–785.

reasons efficacious. Indeed, these experiments may go further and show that the subject (after acting) tends to diagnose his action as having been based on a preference (and he attributes imaginary properties to the option he takes to explain this). But again, these experiments do not show reasons are causally efficacious, they show that after picking an option, people tend to describe what they have done as based on a preference or as a choice. But just because people tend to see themselves making choices as opposed to merely picking, this does not establish that they have chosen. In neither case, then, has the causal efficacy of reasons been established.

In sum: the rationalist says that Oedipus cannot act unless he acts for a reason. Since he does not have more reason to take the road on the right instead of another, in taking it his reason must be causally efficacious. My response is that clearly Oedipus does take the road on the right instead of any other without having a reason. So there must be such a thing as picking. Rationalizing explanation has reached its limit, and an intuitively causal explanation—say, a neurological one—might be invoked to explain various facts (e.g., why he picked the path on the right rather than another) that rationalizing explanation does not and should not purport to give. But although a causal explanation might be invoked here, this does not establish the thesis that reasons are causes any more than the fact that chemistry is needed when, say, biological explanation runs out shows that the original biological explanation was essentially chemical.

Weakness of Will

Suppose Oedipus is weak willed and acts contrary to his own best judgment.[15] Does this not show that primary reasons, together with weighted judgments are not sufficient to (rationally) explain Oedipus's action? For the reasons which were weighted higher according to Oedipus were not, by hypothesis, the reasons for which he acted. Hence we need to introduce causation into the analysis in order to provide an

---

15. This example is due to Nancy Cartwright.

account of rational action. What was irrational about the akratic action was that Oedipus's best judgment did not cause it; whereas when the action is rational it does.

This issue is complicated, and there are a number of ways to tackle the challenge posed above. Davidson's own account of akrasia is rather too complicated to go into here, but in the end it does not support the suggestion that introducing causation is necessary (nor does he take it to be); on the contrary, I think it is his causal account of reasons that renders irrationality paradoxical for him.[16] In any case, this example collapses into those already considered. Let us accept that Oedipus assessed his situation rationally and was "overcome" by weakness or a recalcitrant desire. What does this amount to other than the fact that 1) Oedipus's better judgment does not (rationally) explain his (akratic) act; and 2) Oedipus's other reasons (which were not deemed his best) do rationalize his akratic act? Why must the reasons that rationalize his action also be causes? Perhaps an intuitively causal explanation (hormone imbalance, stress, etc.) can be invoked to explain why Oedipus does not act in accordance with his better judgment. But, as in the case of Buridan's Ass where an intuitively causal account could be brought in to supplement explanation, the hypothesis does not establish that Oedipus's reasons which were deemed inferior by his own judgment were "causally efficacious"—that is, they need not have any more explanatory power than they had in the first place in merely providing (primary) reasons for his action. Alternatively, an intuitively rational explanation might be invoked (e.g., there were two or more internally consistent but intra-inconsistent subsystems competing with each other: one which wanted to act on Oedipal desires, the other which was appalled by these desires). If so, as with overriding reasons, this suggests that the present rationalizing explanation is insufficient and that a more complex justificatory machinery (using concepts of subagency, etc.) is needed. But again, this does not show the necessity of positing a causal relation between reason and action.

16. I address these points in "De-Individualizing Norms of Rationality" (chapter 1) and in n32, below.

Summary

I have considered the appeal to intuition that Davidson makes in "Actions, Reasons, and Causes," to the effect that something essential has been left out of explanation if justification requires only that the agent has primary reasons which rationalize an action, on the grounds that "a person can have a reason for an action, and perform the action, and yet this reason not be the reason why he did it." Five ways of filling out this intuition have been considered. The reasons which are not sufficient to (rationally) explain might be: 1) external; 2) overridden by other reasons or simply not weighted highly enough; 3) inconsistent with what the agent is disposed to say about his own actions; 4) equally balanced with other reasons and thus insufficient to determine a choice between them; and 5) countenanced by the agent's best judgment but not acted upon due to weakness. The first and third cases fail to show the explanatory insufficiency of primary reasons. The second shows the insufficiency of primary reasons, but not the insufficiency of a more complex model of deliberation that takes into account values and weighted judgments; the fourth shows the "insufficiency" of such a model only by forcing it to explain something beyond its purview; the fifth reduces to the second and fourth ones: if it does show the insufficiency of rationalizing explanation, this cannot be corrected by construing reasons as causes.

So much, then, for the appeal to intuition. The "something that is missing" from the primary reason account is, if anything, a more complex interpretation that would include competing reasons, values, judgments manifesting choices, and so on, and no doubt this interpretation will be dependent on what the agent is disposed to say about his own actions. But all of this is perfectly consistent with our talking in an ontologically neutral way about ascriptive "because" statements containing mental predicates which are justified, in turn, by other "because" statements. For example, we might justify attributing reasons having to do with self-defense to Oedipus in part because he believed the old man was trying to kill him. The attribution of this belief, in turn, might be justified because Oedipus saw the old man attack him with a stick, and felt a sharp pain. And so on. But these latter "because" statements are of the type we are trying to understand: without additional argument their use in interpretation cannot be taken to carry with it any substantive ontological commitment (one, for example, that Davidson argues is implied in singular causal statements).[17]

Even if, in justifying particular attributions, we eventually find a "because" statement that is properly explicated by postulating a "real" causal relation between two (perhaps, at bottom, physical) particulars, this in and of itself does not provide any reason to suppose that other, higher-order justificatory statements (using mental concepts) are best construed as describing such a causal relation. Nor, a fortiori, does it provide any reason to think that the original reason statement ("He killed him because he wanted to defend himself") describes one.

Similarly, the idea that certain reasons carry more weight than others might simply be glossed by saying that there may be higher-order justifications one might give for preferring ascriptions of certain reasons over others, even though both reasons are attributable to the agent, and both would serve to rationalize, or provide some sort of justification for, the action. The higher-order story might involve showing why certain reasons, although they provide a rationalization of the action, are not rationally consistent with other values, judgments, intentions, etc., which are also attributable to the agent. But again, nothing in this implies the necessity of viewing the reasons as causally efficacious.

In brief, the kind of explanation afforded here is consistent with a view about mental discourse that takes the attribution of propositional attitudes to involve the introduction of standards in accordance with which a person's statements and actions can be calibrated and, to this extent, explained. But this may well be a discourse in which the attributions of propositional attitudes carry no ontological or explanatory commitment of the kind implied by physicalism.[18]

---

17. I discuss this further in chapters 6, 7 and 8.

18. The kind of explanatory model I have in mind here—one that explains by calibrating the explanandum with respect to some standard—needs further exploration. In particular, we should not insist the standard be practice-independent (as we imagine laws of nature or empirical regularities to be). The role of a standard is explored throughout Wittgenstein's work. For Davidson's own discussion of the idea of belief-as-standard, see

The Motivation for Introducing Causation

3. I HAVE CONSIDERED Davidson's arguments purporting to establish that introducing a causal element into an account of reason-explanation is necessary, and I have argued that nothing is missing that can be provided by construing reasons as causes. I suspect that there is an underlying motivation in Davidson's introducing causation into his account, which perhaps can be gestured at by something like the following intuition. It might be thought that in a proper explanation, the explanans must necessitate the explanandum. This intuition would make the idea that there is something missing from rationalizing explanation understandable. What is missing would be some sort of determinate relation between—as it would have to be glossed—events or states picked out by predicates describing reasons (or best judgment) and action. In introducing causation into the analysis, Davidson would presumably be providing the missing link by providing a (causally) sufficient condition for the occurrence of the action.[19] That this is Davidson's motivation is borne out by the discussion above, where I tried to show that the only contender for "something missing" from the dressed-up rationalizing explanation is—in the Buridan's Ass and akrasia cases—that weighted reasons should determine the occurrence of a particular action. This was the only contender, though in the end, I suggested that insisting on such a relation was too strong a requirement to place on rationalizing explanation.

"What is Present to the Mind?," (1989), *Subjective, Intersubjective, Objective* (Oxford: Clarendon Press, 2001), 53–68, in which he cites Paul Churchland, *Scientific Realism and the Plasticity of Mind* (Cambridge: Cambridge University Press, 1979).

19. Note that although it is widely accepted, it is not uncontroversial, that citing a cause gives a necessitating condition for the occurrence of the event to be explained. See, for example, G. E. M. Anscombe, "Causality and Determination," (1971), in *Causation,* Ernest Sosa and Michael Tooley, eds. (Oxford: Oxford University Press, 1993) for arguments to the contrary. Perhaps another way to explain Davidson's inclusion of causation in his analysis of rationalizing explanation is to see it as motivated by a realist conception of explanation: that is, "the view that an explanation is correct or accurate in virtue of there obtaining 'in the real world' a certain determinate relationship between the explanandum and what is adduced as an explanation of it." Kim, "The Myth of Nonreductive Materialism," 94.

Indeed, Davidson's introduction of causation might be motivated, in part, by an attraction to Hempel's "covering-law" model, which would be a straightforward way to secure a determinate relation between the explanans and explanandum (the explanans would now be construed as consisting of a law, which, together with a singular sentence, logically implies a singular conclusion). Note, however, that introducing causation into the analysis may enable one to conform to such a model of explanatory subsumption, given a host of other assumptions, but it is not necessary. To say that explanation should be modeled as a deductively valid argument is, in itself, neutral on the status of the major premise exhibiting a general relation between types of reason and types of action. It might be, as Hempel thought, an empirical law of psychology (so that psychological explanation fits into a standard "covering-law" model of explanation).[20] Or it might be, as von Wright thought, a statement expressing a conceptual truth about the relation between reasons and actions.[21] The first model would suit an empiricist, but even the conceptual model would suit a certain kind of realist, if the rules were thought to reflect a determinate structure, which (once our conceptual apparatus got going) transcends our epistemic abilities to grasp it.[22]

Davidson is attracted to Hempel's idea that the major premise of the requisite explanation will be an empirical law, but he rejects the idea that the statements couched in a psychological vocabulary could ever be molded into a law of this kind.[23] Davidson is right to think that psychological statements can never be formed into a law—I shall discuss why

20. Carl Hempel, "Rational Action," *Proceedings and Addresses of the American Philosophical Association* (1962): 5–23.

21. Georg H. von Wright, *Explanation and Understanding* (Ithaca, NY: Cornell University Press, 1971). This way of characterizing the model is in Kim, 'Self-Understanding and Rationalizing Explanation," *Philosophia Naturalis* 21 (1984): 309–320.

22. Henceforth, I shall restrict my discussion to the empirical version, but note that many of the arguments of the following section work against the conceptual model as well.

23. Consider: "That a complex causal story goes with a reason-explanation of action doesn't show that laws aren't needed; if Hempel is right, 'causal explanation is a special type of nomological-deductive explanation'"; and "suppose, as I have been, that Hempel is right when he claims that every explanation states or implies an empirical generaliza-

this is so in the next section. But instead of using this fact to abjure a subsumptive model of explanation (which is the real culprit here), and rethinking his positive case for construing reasons as causes, Davidson's doctrine of anomalous monism presents an attempt to hold on to the belief in mental causation by identifying mental and physical particulars. His doctrine of anomalous monism attempts a reconciliation of, on the one hand, the thought that causation requires nomic subsumption and, on the other, the thought that mental phenomena are anomalous. If mental and physical particulars are identical, the story goes, there will be no problem understanding psychophysical causation, since causation is a "metaphysical" phenomenon—a "real" relation between events, whatever properties are referred to in their descriptions. But instantiating a law is a property of events that is not independent of the properties referred to in their descriptions. So the nomological character of causation and the anomalism of the mental are consistent.

This solution, however, does not work; indeed many have recently noted a serious tension between Davidson's anomalous monism and the causal/explanatory role of the mental.[24] I would like to add to this

---

tion. . . ." Donald Davidson, "Hempel on Explaining Action," (1976), in *Essays on Actions and Events,* 265 and 273, respectively.

24. See, for example, Frederick Stoutland, "The Causation of Behavior," in *Essays on Wittgenstein in Honor of G. H. von Wright, Acta Philosophica Fennica* 28, 1–3 (1976): 286–325; Ted Honderich, "The Argument for Anomalous Monism," *Analysis* 42, 1 (1982): 59–64, and "Smith and the Champion of Mauve," *Analysis* 44, 2 (1984): 86–89; Jaegwon Kim, "Epiphenomenal and Supervenient Causation," (1984), and other essays in *Supervenience and Mind* and "Self-Understanding and Rationalizing Explanation"; Dagfinn Follesdal, "Causation and Explanation: A Problem in Davidson's View on Action and Mind, in *Actions and Events: Perspectives on the Philosophy of Donald Davidson,* Ernest Lepore and Brian McLaughlin, eds. (Oxford: Basil Blackwell, 1985); Mark Johnston, "Why Having a Mind Matters," in Lepore and McLaughlin, *Actions and Events: Perspectives on the Philosophy of Donald Davidson,* 408–426; Ernest Lepore and Barry Loewer, "Mind Matters," *Journal of Philosophy* 84 (1987): 630–641; Jerry Fodor, "Making Mind Matter More," *Journal of Philosophy* 84, 11 (1987): 642, and "You Can Fool Some of the People All of the Time, Everything Else Being Equal; Hedged Laws and Psychological Explanations," *Mind* 100 (1991): 19–34; Stephen Schiffer, "Ceteris Paribus Laws," *Mind* 100 (1991): 1–17; Tyler Burge, "Individuation and Causation in Psychology," *Pacific Philosophical Quarterly* 70 (1989): 303–322; and John McDowell, "Functionalism and Anomalous Monism," (1985), *Mind, Value, and Reality* (Cambridge, MA: Harvard University

growing dissension by arguing from a somewhat different angle. Keeping fixed the thesis of the anomalism of psychology (and the psychophysical anomalism on which it depends), we can see that once we understand how the "reconciling" anomalous monism works, whatever prior, independent motivation we might have had for adopting the other two recalcitrant theses is undermined. In particular, once we see how the reconciliation works, the belief in mental causation cannot get any support from the idea that a causal relation between reason and action is required for reason-explanation.

To see this, consider the following: Let us assume that the nomological requirement on causation is intimately tied to something like a Hempelian covering-law model of explanation and nomic subsumption is nomic implication between appropriate event descriptions.[25] But then Davidson's move to go "extensional" with causation, but remain "intensional" with laws, robs us of the explanatory connection between the laws and the particular events that the laws putatively subsume. This is because the "appropriate" event description will surely be one in which the implication goes through—namely, one in which the events referred to in the singular sentence of the minor premise are picked out by the same predicates as the event (types) referred to in the laws set out in the major premise. But if the laws are described by physical predicates and the events that stand in a causal relation are picked out by mental predicates and, as his psychophysical anomalism implies, there are no bridge laws connecting the two domains, then it is difficult to see how the existence of laws can play the requisite role of "subsumption." In any case, laws of this sort certainly do not facilitate explanation, thus there seems to be no good reason to believe in the requirement of nomologicality.[26] I have argued at the outset that intro-

---

Press, 1998), 325–340. See part 1 of John Heil and Alfred Mele, eds., *Mental Causation* (Oxford: Clarendon Press, 1993) for essays on these tensions in Davidson.

25. See Kim, "Causation, Nomic Subsumption, and the Concept of Event," (1973), in *Supervenience and Mind* for a general discussion on nomologicality and causation. This assumption will be discharged below.

26. Suppose that we give up attempting to couch the subsumptive relation between laws and events in a logico-descriptive, or nomic-implicational approach, and attempt instead

ducing causation is not obviously necessary for explanation even when we remained relatively agnostic about what causation entails. But if we were to accept certain assumptions about what causation does entail (nomologicality in this instance) then this gives us a positive reason to reject the introduction of causation into an account of reason-giving explanation.

One can get a better picture of the fault lines threatening the position by making explicit the metaphysics to which Davidson's anomalous monism is committed, and asking whether or to what extent such a picture is required by, or even happily coexists with, reason-explanation. Consider, for example, Davidson's claim that reason-explanation works by attributing dispositions to an agent, which he thinks should be construed realistically.[27] Now, how do explanations involving the attributions of such dispositions work? The answer seems to be that the existence of rough psychological generalizations serves as an epistemic promissory note, as it were, for the existence of strict laws couched in physical terms.[28] But what reason is there for believing this? And

to find a law-like correlation between generic events to sustain a particular causal relation between two individual events. Are we any better off? It is difficult to see how, since, as I shall discuss below in the text, Davidson's anomalous monism leaves us in the dark when it comes to saying which types of physical events must be lawfully correlated in order to sustain some particular causal relation between a mental event (token) and its effect.

27. Davidson seems committed to the realist construal when he says: "Explaining why something dissolved by reference to its solubility is not high science, but it isn't empty either, for solubility implies not only a generalization, but also the existence of a causal factor which accounts for the disposition: there is something about a soluble cube of sugar that causes it to dissolve under certain conditions." Davidson, "Hempel on Explaining Action," 274. The case is similar, according to Davidson, with beliefs and desires. Note that the close links between realism and causation should not be passed over lightly: it is generally conceded that causation buys you realism; that is, if Davidson is successful in showing that reasons are causes then he has, according to many, vindicated the real, distinct nature of reasons in particular, and, if the arguments generalize, to mental phenomena.

28. "[The importance of psychological generalizations] lies mainly in the support they lend singular causal claims and related explanations of particular events. The support derives from the fact that such a generalization, however crude and vague, may provide good reason to believe that underlying the particular case there is a regularity that could be formulated sharply and without caveat." Donald Davidson, "Mental Events," (1970), *Essays on Actions and Events,* 219.

even if it were true, what explanatory help can it be? Without determinate relations between the properties figuring in these different laws (again, something Davidson's anomalism commits him to denying), one might wonder how psychological generalizations could provide a guarantee of the existence of strict physical laws and even if they do, but we cannot access them by type-type correlations, how can the mere existence of these physical laws render events picked out in a mental vocabulary explanatory?

Again, to put the point somewhat differently, consider that according to Davidson's monism, there is one event—call it *c*—which is appropriately described by a mental predicate (or has a mental property) and by a physical predicate (or has a physical property) and this event *c* causes another event *e*.[29] The events that stand in this causal relation are subsumed by some physical law: that is, there is a regular correlation between events of type *C* and events of type *E* which holds between (the instantiation of) certain (types of) physical properties (or there is a regularity which is expressible by a certain physical description). But what any of this has to do with our original reason-attribution is left utterly mysterious since we have no idea how to identify the original event *c* as one that is apt both for the appropriate mental and physical predicates/properties to begin with.[30]

---

29. Both truth (justification) and relevance (i.e., for explanatory purposes) must somehow figure in the "appropriateness" condition.

30. Compare: "What is so unsatisfying about [Davidson's anomalous monism] as an account of the mind-body relation is the fact that it says nothing about the relation between mental and physical properties; the only positive thing it says about that relationship is that mental and physical properties are co-instantiated in objects and events." Kim, "Postscripts on Mental Causation," *Supervenience and Mind,* 365. Kim's favored alternative to anomalous monism (or token physicalism)—multiple-type physicalism—also attempts to solve the problem of mental causation by countenancing only one event which instantiates both a mental property and its physical "realizer." On this picture, mental properties map onto one or other of its realizing physical properties; each instance of a mental property can be reduced, though they cannot be reduced all at once. This view—which, according to Kim is probably the only way to save mental causation—is subject to many of the worries expressed in the text above: one's motives for countenancing mental causation of this sort are far removed from the original arguments that seemed to require such a relation for explanatory purposes.

Many recent commentators on Davidson have judged that his doctrine of anomalous monism threatens the causal power of mental properties, thus leaving them subject to the status of "epiphenomena" and thereby reintroducing a kind of dualism that a respectable physicalist ought to shun.[31] Some of these commentators have tried to stave off the threat of mental epiphenomenalism by claiming that causation only requires subsumption under "hedged laws": true, counterfactual supporting generalizations, hedged by ceteris paribus clauses. But this move also fails: mental predicates resist capture in any model of explanatory subsumption. In the next section, I shall argue that there are no such things as rational-psychological hedged laws: what appear to be law-like generalizations are either not true because incomplete, or if the ceteris paribus clauses are filled in, then vacuous. If this is so, one must reconsider the motivation for linking causation with subsumption in the first place, or, at least with the general intuition that in a proper explanation, the explanans necessitates the explanandum. If causal explanation is best construed on this model or on one of its variants—i.e., as providing a necessitating condition for the event to be explained—then, as I have suggested, this gives us a positive reason to reject the idea that the explanatory power of reasons is causal.

In the next section I would like elaborate on what is worrying about adopting nomologicality—or any version of explanatory subsumption—as a model for psychological explanation.

Irrationality and Practices

4. I HAVE ARGUED that when a person acts and has reasons that rationalize his action, we do not necessarily need more from rationalizing explanation. We might be able to add weighted reasons and values

31. I think that much of the defense of Davidson (focused on the propriety/impropriety of causally relevant properties) misses the point that Davidson's view of explanation seems to commit him to there being subsuming laws, or at least law-like generalizations, couched in the vocabulary of the causal statements. It is difficult to see how he will ever reconcile this with the anomalism of the mental.

to our rationalizing explanation, but this "ascension" at the level of interpretation need not reflect causal processes occurring at the "object level." If in positing a causal relation between reason and action one is hoping to ground a determinate connection between (weighted) reason and the action it rationally explains (or their associated descriptions), one's hope is misguided: the possibility of irrationality (or of merely picking and not choosing) would be ruled out if this strong relation obtained.[32] To see this, let us consider how we might attempt to formulate the psychological generalizations used to supply the major premise in a covering-law model of explanation.[33]

Most psychological generalizations will have defeating conditions and so will need to be constructed with complex ceteris paribus clauses.

---

32. Davidson thinks that intentions or singular value judgments determine the actions they prescribe (indeed, it is his account of intention that provides part of the evidence for thinking he aspires to a covering-law model of explanation). But this leaves him unable to cope with irrational phenomena, so in "How is Weakness of the Will Possible?" he argues that although intentions determine action, all things considered judgments do not, thus creating a "logical gap" wherein the irrational desires may gain a foothold. In "De-Individualizing Norms of Rationality" (chapter 1) I argue that Davidson's account of akrasia suffers precisely because he wants to allow irrationality while his underlying motivation for introducing causation into the analysis makes it impossible.

33. In "Ceteris Paribus Laws," Schiffer also stresses the impossibility of psychological laws. He challenges Fodor, who subscribes to the possibility of hedged laws in psychology, to find a completion for the ceteris paribus clause of a statement ostensibly expressing such a law. This ceteris paribus clause would, according to Schiffer, have to accommodate both our intuitions about the basicness of physics and about the causal relevance of the mental properties referred to in the law. The challenge is leveled specifically against a supervenience relation between the mental properties and their physical realizers to do the requisite work: the possible multiplicity and diversity of the physical "realizers" are thought to be intuitively too disparate and various to determine the causal relevance (sufficiency) of the mental properties they determine. This is a version of the standard, epiphenomenalist objections to Davidson's anomalous monism—though aimed specifically at the possibility of completing ceteris paribus clauses such as to get the right relation between physical and mental defeators of the law. Schiffer is right to worry about such completions but the trouble arises at a more basic level: such laws are impossibly circular on the one hand, and on the other, they leave unaccommodated a whole domain of (irrational) phenomena which falls within the purview of reason-explanation. This objection holds for generalizations described in a mental vocabulary before (quasi-) reduction is even attempted.

One particular condition needs to be ruled out right away: that the individual is irrational. The problem is that we allow violations of practically everything that constitutes a candidate for a psychological law. Consider, for example, the generalizations that Paul Churchland cites in support of his claim that folk psychology is a theory:[34]

i) (x) (p) [(x fears that p) → (x desires that ~p)];
ii) (x) (p) [((x hopes that p) & (x discovers that p)) → (x is pleased that p)];
iii) (x) (p) (q) [((x believes that p) & (x believes that (if p then q))) → (barring confusion, distraction, etc., x believes that q)];
iv) (x) (p) (q) [((x desires that p) & (x believes that (if q then p)) & (x is able to bring it about that q)) → (barring conflicting desires or preferred strategies, x brings it about that q)].

The first two are at least misleading, if not false, once we take into account normal ambivalence. (Try substituting for p: "y loves x" or "x is pregnant.") The third is false if we admit the possibility of self-deception, and so is the fourth if we admit the possibility of akrasia.

One might attempt to dismiss this problem by concentrating on psychological generalizations concerning those we assume to be ideally rational. But this would amount to a serious overgeneralization. The problem then would be to explain why it is that those of us who so often fall foul of ideal rationality are nonetheless relatively easy to explain psychologically even when we are in the midst of our irrational moments. Akrasia is an example of irrational intentional action. To the extent that it is a species of intentional action, it is explicable by citing a reason. But how does this reason-explanation work? In virtue of the fact that (overriding) reasons by and large necessitate (or even determine the probability of) actions that they rationalize? This cannot be: we are examining a case (the akratic one) where this relation has failed but the action is rationalizable and to that extent explicable

34. Paul Churchland, "Eliminative Materialism and the Propositional Attitudes," *Journal of Philosophy* 77, 2 (1981): 67–90.

nonetheless. So, in order to accommodate these cases we would need to specify the defeating conditions in a way that does not overgeneralize; but then the worry is that these more precisely defined "laws" will rather quickly reduce to: Given certain beliefs and desires, a person will Ø unless he does not.

This is a version of what is currently dubbed the "normativity" problem for (mental or linguistic) content rearing its head in action theory.[35] In thought, perception, language, and action, mistakes or errors can occur and yet still not impugn the identity of the mistaken item as a thought, percept, meaningful utterance, or action. My erroneous beliefs are still beliefs about something, my akratic actions are still intentional actions, my malapropisms still meaningful. But any proposed relation between the representation and what is represented that forges the connection too close seems unable to account for errors; and yet any relation that slackens the requirement to accommodate errors seems to fall short of providing the kind of relation that those offering substantive, constitutive accounts of intentional phenomena are hoping to provide: namely, a set of necessary and sufficient conditions for a representation to have content in terms of what it represents. Although the normativity problem has been framed in terms of the putative individuative conditions for some representation, it has ramifications for the view under consideration regarding the explanatory relation between reasons and action. Construe the representation as an action: what it represents as the reasons that rationalize it. The explanatory connection between reason (what is represented) and action (representation) cannot be too close—reason cannot necessitate or determine action—or irrationality (or picking, as opposed to choosing) will be impossible.

Most of today's literature on the subject is particularly concerned to provide "natural" reductions of meaning and intention. What it is for a belief to be about something, for example, or for an expression to mean something, has to be set out in terms of necessary and

---

35. For various attempts to describe the normativity problem see the Introduction in Ruth Millikan, *White Queen Psychology;* Paul Boghossian, "The Rule-Following Considerations," *Mind* 98 (1989): 507–549; and Fodor, *A Theory of Content.* My formulation of the problem is rather different, as are my conclusions.

sufficient conditions in which no intentional or semantical notions figure. Indeed, most attempts are thought to fall foul of the naturalistic requirement.[36] But, if I am right, the normativity problem will plague anyone attempting such an account, whether it is naturally specified or not. In short, if intentional phenomena are either logically or causally determined by any particular norm (of reason, meaning, or perception) that governs them, then error internal to these practices (irrationality, malapropisms, or misperceptions) cannot be diagnosed as a violation of this norm. Perhaps what is required for normative phenomena (representations to be about something) is merely that a sufficient number of norms are satisfied; but again, no one norm will be necessary if violation of this norm, or if this particular species of irrationality, is to be possible.[37]

I suggest that we give up the conviction that to be explanatory, reasons have to determine (or determine the probability of) actions that they merely prescribe. And if this argument strategy iterates so that particular attributions of reasons are not determined, but merely prescribed, then we are left with a substantial indeterminacy in what counts as an apt ascription.[38]

36. See Boghossian, ibid., 540, where it is clear that his arguments (basically from the holistic character of belief fixation) against a dispositional theory of semantic or intentional content tell against any attempt to state necessary and sufficient conditions for being a belief with a certain content—not only against naturalistically specified conditions.

37. Compare this with Wittgenstein's family resemblance view of concepts.

38. One needs to resist the temptation to argue that because psychological terms figure in empirical generalizations that this is all we need for psychological laws, and the existence of these laws will be all that is required for an empirical investigation of psychology. I have tried to suggest that the way in which putative psychological terms figure in explanations of this kind is quite independent of their role in rationalizing explanations (and arguably this demotes their status as genuine psychological concepts). It would be interesting to compare this with Davidson's own discussion of why there cannot be psychological laws. He gestures, I think, in the direction I take when he alludes, in "How is Weakness of the Will Possible?," to Quine's indeterminacy thesis of translation and to the heteronomic quality of any putative psychological law. (One would be changing the subject and robbing the psychological of its essential, constitutive properties if one were to change vocabularies in order to turn a hedged, psychological generalization into a strict law.) I do not think that it has been generally recognized why the indeterminacy of intentional properties is essential (i.e., that it allows irrationality or norm violation).

While the constitutivist is looking for empirical regularities, conceptual truths, or even judgments to ground reasons I suggest instead that we take Wittgenstein's advice not to think but to look: look at the way reason-explanation works, by investigating the practices in which reason-concepts figure.[39] Look at the way reasons are ascribed, look at the extent to which these ascriptions are subject to correction, and look at what the participants of the practice accept as reasonable justification for them. The justifications on offer for adopting certain ascriptions rather than others are underdetermined by the norms that govern the practices of reason-explanation—that, I take it, is the main lesson of Wittgenstein's "rule-following considerations." Perhaps reflection on these practices will show that the indeterminacy in the rules that is essential to accommodate the possibility of norm-violation makes room for a subject-dependent element (pickings, optings, reactions) which in turn might provide the only way we have of "accounting" for ascriptions which seem determinately fixed. Taking a look within our practices and leaving the logical space for justification to "run out" or "reach bedrock" will not satisfy someone hoping to offer a constitutive account of the intentional phenomena—but it is nevertheless a legitimate way—perhaps the only way—of investigating the mental.[40]

Conclusion

5. WHERE DOES THIS leave us with mental causation? Are reasons causes? Our ordinary conception of causation probably involves some admixture of rational relations and empirical relations.[41] What I have

---

39. See Wright, "Wittgenstein's Rule-Following Considerations and the Central Project of Theoretical Linguistics" (1989), *Rails to Infinity: Essays on Themes from Wittgenstein's* Philosophical Investigations (Cambridge MA: Harvard University Press, 2001), 170–213.

40. I say more about how such an investigation might proceed in "Investigating Cultures: A Critique of Cognitive Anthropology," *Journal of the Royal Institute for Anthropological Studies* 4, 4 1998): 669–688.

41. Thus, agent-based analyses of causation, e.g., Douglas Gasking, "Causation and Recipes," *Mind* 64, 256 (1955): 479–487.

attempted to show is that the explanatory force of mere rationalizations does not need support from the putative existence of determinate (or for that matter, statistical) regularities—empirical or otherwise. Perhaps there are good reasons for restricting causal relations to those that do involve determinate (statistical), empirical regularities or at least to explanatory domains that aspire to "discover" them, as the physical sciences do.[42] Doing so may enable us to preserve a notion of causal realism (and explanatory realism) for certain practices (our scientific ones) and, at the same time, allow us to decline from taking a metaphysical stance on issues like the reality or irreality of mental properties. It is, after all, the assimilation of explanatory paradigms that forces this issue, and it is the assimilation of explanatory paradigms that has made the problem of the mental so intractable.

Consider, in closing, the following passages from three recent pieces in philosophy of mind. The first consists in the closing lines of an article by Terence Horgan, which describes in detail the failure of various attempts to formulate a supervenience relation of the mental upon the physical that can satisfy the demands of a nonreductivist physicalist position.

> Explaining ontological supervenience relations in a materialistically acceptable way looks to be a very daunting task, *whose difficulty suggests the need for materialists to consider seriously the prospects for preservative irrealism about much of our higher-order discourse.* It is not easy formulating a metaphysical position that meets the demands of a material world; there is still a lot of philosophical work to do.[43]

42. This last qualification is needed, since one might argue that even in the physical sciences, we fall short of having true, determinate laws to constitute our theories. Even so, there is an important difference between the physical and the mental domain in this regard. Physical sciences at least aspire toward discovering determinate, empirical regularities. This is part and parcel of the objectivity that is sought in scientific investigations. But I have given some reasons to think that psychology (and investigations of other norm-governed practices) cannot aspire to this kind of objectivity in its investigations.

43. Terence Horgan, "From Supervenience to Superdupervenience: Meeting the Demands of a Material World," *Mind* 102 (1993): 555–583. My italics.

In the next, Jaegwon Kim presents a very strong and sustained attack against the possibility of nonreductive physicalism; rather than embracing an irrealist position, he somewhat tentatively concludes that a localized kind of type physicalism must be true:

> All of this assumes that the mental is "physically realized." Is that true? Should we think of the psychophysical type-type relationship in terms of "realization"? . . . Ultimately we are likely to face the following choice: either embrace the realization view and save mental causation, or insist on the unique and distinctive status of mental properties . . . but be prepared to give them up as causal powers. The paradoxical thing about this is that the choice offered may only be an illusion of a choice, for the two options may in the end collapse into one. If you choose the former, you may lose what makes the mental distinctively mental; and what good is it, one might ask, if you save mental causation but end up losing mentality in the process? . . . If you choose the latter, again you may lose the mental, for what good is something that is causally impotent? Why should we bother to save belief and desire . . . if their presence or absence makes no difference to anything else and we can't use them to explain anything? Being real and having causal powers go hand in hand. . . . We therefore seem to be up against a dead end. Perhaps, that is what's really so intractable about the problem of the mind.[44]

Finally, let us consider an excerpt from the last paragraph of Paul Boghossian's discussion of Kripke's interpretation of Wittgenstein's discussion of rule-following. Based on the failures of various reductionist strategies, and (what Boghossian argues is) the incoherence of content

44. Kim, "Postscripts on Mental Causation," (1993), *Supervenience and Mind,* 366–367. It is somewhat surprising that Kim decides not to mention the promise of exploring other, non-causal models of mental explanation in the conclusion of this collection of essays; he, himself, suggests it as a promising line of inquiry in "Self-Understanding and Rationalizing Explanation" and "Mechanism, Purpose, and Explanatory Exclusion," *Philosophical Perspectives* 3 (1989): 103n4.

irrealism, Boghossian is tempted toward an "anti-reductivist" position, but raises for it the following worry:

> Finally, though, there is the question of mental causation: how are we to reconcile anti-reductionism about content properties with a satisfying conception of their causal efficacy? It is a view long associated with Wittgenstein himself, of course, that propositional attitude explanations are not causal explanations. But, whether or not the view was Wittgenstein's, *it has justifiably few adherents today.* As Davidson showed, if propositional attitude explanations are to rationalize behaviour at all, then they must do so by causing it. . . . But . . . how is an antireductionist about content properties to accord them a genuine causal role without committing himself, implausibly, to the essential incompleteness of physics?
>
> This, I believe, is the single greatest difficulty for an anti-reductionist conception of content. It may be that it will eventually prove its undoing. But the subject is relatively unexplored, and much interesting work remains to be done.[45]

I think it is clear from looking at these passages that the time has come, pace Boghossian's remarks emphasized above, to reopen the debate as to whether reasons are causes. I have suggested that answering this question might, in the end, depend on the extent to which causation, realism, and various aspects of the model of explanatory subsumption are, or ought to be, coextensive. Kim has argued that there are good reasons to require of anyone wanting to be a realist about causation (one who believes that there is a determinate objective fact of the matter about the causal history of a given event) that he aspire to only one explanation for a single explanandum (where, for example, rationalizing and neurophysiological explanation may not each provide complete and independent explanations.)[46] One could, perhaps, use this point to urge that something like the Hempelian covering-

45. Boghossian, "The Rule-Following Considerations," 549. Italics mine.

46. Kim, "Mechanism, Purpose, and Explanatory Exclusion."

law model is the ideal of scientific explanation, and correlatively urge that psychological explanation is not best construed as conforming to it, nor as best construed as causal. In any case, in arguing—against Davidson's claims to the contrary—that there is nothing missing from rationalizing explanation that requires construing reasons as causes, I hope to have prepared some of the groundwork for the redeployment of this widely neglected position.

CHAPTER SIX

Reason-Explanation and the Contents of the Mind

1. WHEN WE ARE puzzled why someone acts as she does, our puzzlement often disappears when we learn more about the circumstances in which the action takes place. Why did the woman run out of the building? The building was on fire.[1]

In being satisfied with learning about the circumstances we do not necessarily want or need to know anything more—anything more particular, say, about the agent's point of view or state of mind. This raises the following question. In asking for explanations of actions, and in being satisfied when we learn more about the circumstances in which the action occurred, have we not learned the reason for the action? Many philosophers think not: we have not been given a reason-explanation until we learn something more—something particular about the agent. We need to know her reason for acting as she does. According to some philosophers this requires knowing the reason that moved or motivated the agent to act. Knowing this is usually thought

1. By "the circumstances" I mean facts about the situation that are not hidden in the way I shall go on to describe.

to involve either knowing the reason the agent endorses, the reason that causes the action, or some combination of the two. These, in turn, are sometimes thought to rest on the occurrence of something "hidden": a particular thought, inner saying, or some other, introspectively unavailable, internal event.

Here is a slightly different route to a more specific version of the conclusion that statements containing mental concepts are made true by the existence of something hidden or inner. We describe people using mental concepts. These concepts pick out mental states. We (at least implicitly) refer to a person's mental states in giving reason-explanations of her actions. This kind of explanation involves causation. Minimally, causation is a relation between two temporally distinguishable events. (Less minimally, the relation is subsumed by laws, which it is the business of the empirical sciences to discover.) Therefore the mental states referred to by mental concepts in reason-explanations have to be the kind of state or occurrence that can enter—either directly or indirectly—into genuine causal relations. Discrete physical states or events of the organism are the only plausible contender for either fulfilling or grounding this role. Therefore the mental concepts that are used to explain and understand a person must refer either directly to physical states of the organism (in virtue of being identical to them) or indirectly to physical states of the organism (via a functional or interpretational specification of them).

These are slightly different paths to the conclusion that mental concepts home in on something hidden or inner. The more specific conclusion of the second argument—that these concepts home in on something at bottom physical—stands behind the current orthodoxy in analytic philosophy of mind and a certain theoretical, though perhaps avoidable, conception of the underpinnings of cognitive science. There are many ways of reaching the conclusion and there are many forms that a resistance to this conclusion might take. We explain individuals and their actions by wielding mental concepts. Although I do not argue for this point here, I am not convinced that this use of mental concepts is best construed as referential. And unless we are willing to give up on what I described as a minimal condition for causation then reasons cannot be construable (in the normal case) as the causes of her actions. I have argued for these points elsewhere.[2] In this paper, I shall develop a different form of resistance by taking a close look at the kinds of considerations we actually use to reach agreement in our ordinary (nonphilosophical and nontheoretical) judgments about a person's reasons for acting. I shall argue that our reason-giving practices offer no support for the conclusion that these concepts are normally used to pick out inner states of the individual that are in some sense hidden from us. They would be so hidden if we were referring to occurrences that were only introspectively available. They would also be so hidden if we were referring either directly or indirectly to physical states of the organism. This line of thinking suggests trouble both for mainstream accounts of what it is to act for reasons and for the current philosophical orthodoxy on the nature of mental states.

The practice of ascribing reasons and the kinds of understanding it brings is familiar, as is the family of expressions that revolves around the notion of "having reasons for acting." These expressions have well-established uses in ordinary (nonphilosophical and nonscientific) discourse. There is no reason to suppose that every time we use the expression "reasons in acting," "reasons for acting," "having reasons," or some cognate expression—even if we can stipulate subtle differences between them—there must be something—some set of characteristics or properties—in common. Nor are there any grounds for supposing that every reason-explanation must conform to the same pattern. Perhaps expressions like these are used in a variety of ways, and the kind of understanding they are marshaled to deliver also varies, with only overlapping or crisscrossing similarities between them. This does not bode well for the prospects of constructing an account of acting for reasons or a theory of reason-explanation. Far worse, though, for the prospect of theory construction is the thought that acting for reasons, or agency, is somehow basic because the notions cannot be elucidated in a way that does not presuppose what is to be explained.[3]

2. In "Why Reasons May Not Be Causes" and "Conceptual Amorphousness, Reasons, and Causes" (chapters 5 and 16).

3. See "Playing the Rule-Following Game" (chapter 3) for a general argument along these lines.

For these reasons and others that will become clear in the course of this chapter, I am deeply pessimistic about the possibility of constructing a theory or an account of acting for reasons. Instead of trying, then, I shall help myself to ordinary notions of what it is to act for reasons, and I shall probe into how we give, accept, and dispute reason-explanations in order show that we are not obliged to accept the philosophical thesis that someone's acting for reasons depends on the existence or occurrence of something hidden or inner which is a candidate springboard for her action. One strand of this argument involves reminding the reader that it is often the case—and this is one thesis of the chapter—that all we need to relieve our puzzlement is a wider view of the context or the circumstances in which the action takes place. In those situations where we do need to enquire about the agent's conception, this does not, in the normal case, lead us to look for hidden, inner events that may be causes of her action. This is the second thesis of the chapter. Finally, even though there are cases in which the agent's introspections, reflections, and deliberations are relevant to our search for reasons, such cases do not lend support to the idea that reasons are hidden or inner causes of action, nor do these cases provide the model for how reason-explanation in general should be understood. On the contrary: there are good reasons to see these cases as special and dependent on the other ones. This is the third thesis and final strand of the argument.

2. THE WOMAN RAN out of the building. Why? It was on fire. Learning more about the circumstances often permits us to understand the action by providing a reason for it without our having to probe more deeply into how it seemed from the agent's point of view. In normal circumstances, that is, the building's being on fire provides the reason for a person's running out of it. But this is not because awareness of facts alone can motivate, move, or cause an agent to act: whether someone was motivated or moved to act by considering these circumstances is a question that normally does not arise.

Learning more about the circumstances may not, of course, relieve our puzzlement why a person acted as she did. Some explanations will have to mention the agent's conception of the circumstances. This will be so when the agent has a misconception that explains her action. Suppose the building was not on fire. The woman might have fled because she thought it was. But it does not follow from the fact that some explanations mention the agent's conception of the circumstances that all explanations do—even, say, when there is no misconception involved.

This last thought might be challenged on the grounds that the circumstances may not tell us all we need to know. Even if we understand why anyone in her situation would run out of a burning building, this may not enable us to understand the action of the one who fled. The building's being on fire will not explain why the woman ran out of the building, if, for example, she did not know that the building was on fire. This is why we need to know how she conceived of her circumstances. And whereas before, the reasons we accepted were in a sense "outside" her or "external" insofar as they allow us to endorse the action as something that makes sense in the circumstances, now it looks as if we are enquiring into the agent's "mental states." And what are these, one might argue, if not something in her mind, and thus, one presumes, her brain? I have suggested that (in the normal case) an appeal to the circumstances is often enough. Now let it be conceded that sometimes we do have to appeal to the agent's conception of the circumstances in order to discover why she acted. But I shall deny that in appealing to how she conceives her situation, we must be homing in on something hidden or inner.

The woman ran out of the building; it was on fire. If circumstances were such that she could not have known the building was on fire, we would have to cast about for a different explanation. Short of special circumstances, I submit, we would tend to accept this as the explanation why she fled. Supposing this is so, then there is a sense in which we have already gained access to her mind by knowing what reasons there are for her to have acted. We have gained access to her mind, that is, insofar as this explanation, in the circumstances, admits a straightforward "logical or rational reconstruction" of her action: she ran out of the building (we might agree to say) because she realized it was on fire. But such a reconstruction does not involve hypothesizing anything hidden or inner. It indicates no more than that we accept the building's being on fire as an explanation why she fled because, *inter alia,* we have no reason to suppose she could not have known it was on fire.

We might also include in this reconstruction a nod to her so-called "pro-attitudes" as well: as long as we have no case for supposing she wanted to burn (and unusual circumstances these would have to be) we might say she realized the building was on fire and she did not want to burn. There is nothing wrong in recasting her reasons for fleeing the building this way, but nor is there any reason to suppose we must do so in order to understand why she left the building—or that such a construction is implicit in our understanding. A person's state of mind can sometimes be read off a pattern of behavior in the circumstances. This is not because we are making inferences to the best explanation by hypothesizing hidden causes, nor because there are analytic entailments between statements containing mental terms and those containing behavioral ones. It is because part of what it is to use mental concepts is to explain an individual's actions in the circumstances.

We might be wrong in our attempts to construct reasons, of course, and it is this possibility that tempts us to think that we must know how the agent, herself, conceived her circumstances. We cannot, for instance, construe the reason why the woman fled the burning building simply by noting that it was on fire if we have independent or competing evidence for saying she did not know this. But "independent evidence," here, does not amount to evidence for the existence of a different hidden, internal state. It is evidence, rather, that the circumstances were not as we first construed them. Once we have a more inclusive view of the circumstances—for example, the woman ran out of the building before the alarm sounded, the fire started on the 12th floor, and she was in the basement—we might have to search again for an explanation that will satisfy us. A shorthand way of indicating our unwillingness to accept this as an explanation why she fled the building would be to reject the explanation on the grounds that she could not have known it was on fire.

Our use of mental concepts in explaining the actions of an individual answers to the same requirement as citing facts about the circumstances in searching for her reasons. That is, we not only cite facts about the circumstances but also attribute intentions, beliefs, and other propositional attitudes to an individual because so doing helps explain her actions. In talking here about what the agent realized or failed to realize, or knew or did not know, then, we have not yet moved away from the practice of reason-explanation. Talk about what she knew or realized is simply in this case a reconstruction of how it makes sense for her to have acted in the circumstances. We do not have, in particular, a more detailed understanding of her mind that is not derived from our quest to understand her reasons for acting.

We have not, of course, preempted all defeating stories we could tell. Suppose we accept that she ran out of the building because it was on fire. Suppose that nothing in her circumstances gives us any reason to suppose she could not have known the building was on fire, so we accept that she realized it was on fire. Thus, we accept the explanation that she ran out of the burning building because she realized it was on fire. But now let us suppose she is a firefighter, and only ran into the building in the first place because she had been called to the scene. Now we really are in need, it seems, of some independent access to her mind that cannot just be read off a description of the action in the circumstances. (This intuition put the nail in the coffin of behaviorism.) So we probe further. Do we hypothesize inner, hidden causes? No, we look in more detail at her circumstances.

Perhaps she could not have put the fire out on her own. This might relieve our puzzlement as to why she fled, unless there is evidence to suppose she thought otherwise. This weak requirement that there be no evidence to suppose she thought otherwise, as we have seen, tends to get turned into the stronger-sounding requirement that she must have realized she could not have put the fire out on her own if this is, indeed, the reason she ran out of the building. Again, there is nothing wrong with acknowledging the weak requirement that there be no evidence to suppose she thought otherwise by saying she must have realized she could not have put the fire out on her own, as long as it is recognized that saying this so far has not introduced something independent into the picture beyond our acceptance of that as the reason which explains her action. Saying she must have realized she could not put out the fire on her own does not signal the occurrence of any inner event: it merely reiterates our acceptance of the fact that she could not have put out the fire on her own as an explanation of why she fled the burning building (together with the fact that we have no evidence or reason to think she thought otherwise).

Perhaps one of the occupants was trapped on the roof and she ran out of the building to fetch a ladder. Finding evidence that she could not have known that someone was trapped on the roof will thwart this description of her (where again, what she knows or does not know amounts not to something hidden inside her but is a simple reconstruction of a wider look at the circumstances). So will a more detailed knowledge of other defeating factors. We cannot explain why she ran out of the building by appealing to her realization that an occupant was trapped on the roof if we have independent reason to suppose she had no thoughts of saving him.

But again, such a weak characterization tends to get transformed in philosophical discussions into the positive requirement that she must in those circumstances have had thoughts of saving him. What does this mean? Surely not that the sentence: "I must save him!" flashes through her mind. For even if such a sentence does flash through her mind she might not mean it, or indeed act upon it. Surely not, *a fortiori,* that a sentence with the same meaning is tokened that belongs to a proto-language of which she was unaware, which tends to engage with various subsystems in a way that disposes her to save lives. For even if such a sentence is tokened, and even if it is so disposed to engage with other subsystems, it might not on any particular occasion engage in the way it is disposed to engage. Nor, even if it does so engage, would this account for the commitments she incurs from having the conviction that she must save him (like the remorse she ought to feel if she does not try). In any case, determining whether a sentence has flashed through her mind, or whether a proto-sentence has been tokened and has engaged in the way it is disposed to engage, plays no role whatsoever in an investigation whether or not she did, on that occasion, have thoughts of saving him. So it is very odd that the occurrence of any such item is thought necessary for a metaphysical, scientific, or so-called "naturalistic" account of what it is to act for reasons.

Perhaps we have independent evidence that she wants the man dead. Then we might reject the explanation that she ran out of the building to fetch a ladder to save him: perhaps she ran out to help ensure his death.

But on what grounds would we agree that she wants the man dead? Is it not the same story all over again? We often use the concept of wanting, just as we use concepts like "realizes," "believes," "knows," and "thinks," not to hypothesize inner, hidden events or causes, but rather to attempt to make sense of the agent and her actions. Coming to see how to endorse the action as making sense in the circumstances is often enough to remove our puzzlement. A prosecutor might begin his case against her simply by noting that she, a firefighter who was equipped to handle the fire, ran out of the burning building even though there was a man trapped on the roof. Her wanting him dead, of course, cannot simply be read off or reconstructed from this strange set of facts. She might have fled the building for another reason. And her wanting him dead is consistent too with her not having fled the building in order to bring about his death. (She might have fled because her own life was in danger.) Her defense counsel will look for another story to explain why she fled the building and independent evidence to back it up. Perhaps she fled in fear—in which case she might be guilty of negligence but not manslaughter or murder. A prosecutor will look for independent evidence that she wants him dead. What would constitute such evidence?

Why, being able to make sense of her failing to save him by reconstructing or describing other aspects of the circumstances! The man might have been the arsonist who started the fire that had killed her brother. This explanation would be withdrawn immediately if our firefighter could not have known that this was the same man, or if she had not known that the death of the firefighter had been the death of her brother, or if she did not know about the death of her brother because she had not known she had a brother. But our independent grounds for describing her in any of these ways will again involve an investigation of the circumstances: she could not have known she had a brother because her mother was away at the time, gave birth to him in secret, and never told anybody about him subsequently.

Even if all this were true, we might be tempted to probe further for an explanation if there were independent evidence to suppose that saving lives was something she lived for and revenge was something she despised. But even this would not, in and of itself, defeat the explanation that she fled the building because she did not want to put out the fire because doing so would have saved the man trapped on the roof and she wanted him dead. Maybe she, like most of us, can be tempted to "act from desires" that she despises. If so, perhaps we will now agree that this is the reason why she fled the burning building. Or perhaps we will not be able to reach agreement about her reasons for fleeing.

We began with the thought that in asking for an explanation of an action, we might be satisfied by learning of the circumstances in which the action took place. When we were not so satisfied, we probed deeper into the circumstances in which the agent found herself. True, we adverted at times to the agent's mind. But our use of intentional concepts, like our description of the wider circumstances, was motivated by what would enable us see her action as reasonable. But not once, in reconstructing how these explanations are offered, rejected, and accepted, have we attempted, or felt the need, to probe into some inner or hidden state of the agent. The use we made of mental concepts to interpret her seemed a way of allowing us to understand the agent's actions in the circumstances.

This characterization of our practice of reason-explanation seems apt for certain kinds of agents—human beings whom we can observe but with whom we cannot engage (like historical figures or figures we are watching from afar). In a much more rudimentary form, this characterization of our practice of reason-explanation is appropriate for animals and small children. When we encounter agents who can communicate with us and answer our questions as to why they act as they do, then our search for reasons becomes more complicated. Such an agent shares with us much more than the primitive ability of animals and infants to move about the world in a way that licenses descriptions of their behavior as rudimentary actions. In being able to communicate with us about her reasons for acting, such an agent shares a common social training in the activities or practices upon which our ability to see an event as an action rests. And this shared social training includes training in discourse about reasons—a discourse that further enriches the whole character of the concepts of reason, action, and agency by, among other things, manifesting the connections between these concepts and those of wanting, intending, believing, hoping, and so on.[4] Someone trained in this discourse is able to make much finer-grained distinctions—and even introduce new possibilities—in talking about her reasons for acting than we may ever be able to do simply by watching her. This may indeed introduce a genuine role for introspection, reflection, and deliberation to play in a characterization of how we search for reasons—at least in some cases.

I shall return to this briefly. But for now I shall emphasize that, important as this skill may be, the one who is able to talk about her reasons for acting is bound by this shared training and the understanding it makes possible. Looking in more detail at how she is so bound will help tell against the idea that in general, a search for reasons involves the search for something hidden or inner: a candidate cause of her action.

3. Suppose we ask the firewoman why she fled the burning building. What would her answer tell us? It depends on what she says. She might claim that she had always sought revenge on the man who was trapped on the roof because he is the arsonist who had started the fire that killed her brother. But in what sort of circumstances will we count this as a confession (as opposed to say, an irrational rave or an attempt to cover up for mere cowardice)? We will count it as a confession only to the extent that we are willing to accept this as the explanation, and accepting this as the explanation will in turn depend on our believing her story. If we are inclined not to believe her story then we will begin, of course, a more in-depth investigation of her circumstances.

To bring this point home, suppose our firefighter said she ran out of the building to get something to eat. She, a firefighter on duty, had been called to the scene, the fire had just started, and she was equipped to put it out. Yet, according to her, she wanted something to eat and so left the building. Pending unusual or strange circumstances, we would

4. For a discussion of this interconnected conceptual landscape, see A. I. Melden, *Free Action* (London: Routledge & Kegan Paul, 1961), chapter 13. The sentence in the text adapts part of what Melden says on page 197.

reject this as an explanation on the grounds that it does not make sense. Suppose she says that it makes perfect sense to her why she would drop everything—even put lives at risk—because she wanted something to eat. We just do not understand what it is like for her when she wants something. Indeed she is right: we do not understand. Assuming no unusual circumstances, we would not only reject her wanting something to eat as a reason and as an explanation for her fleeing the building; we would also begin to doubt the intelligibility of her words—that is, we would begin to doubt that we could make out what she means by saying them here. Not anything someone says will be accepted as the reason why she acted.

There are difficulties in any case in understanding what is meant by "what the agent says." Do we mean what she does in fact say as she acts? She might not have said anything, even to herself. Or perhaps the only words she uttered were in the form of the mantra she learned to recite over and over again in difficult situations. If we cannot find her reasons for acting in what she said to herself as she fled the building, then perhaps we ought to look for her reasons in what she would say about her reasons for acting, if asked. Or do we mean, rather, what she would say if she were to reflect? If she were open to reason in reflecting on why she fled, she would soon come to see that wanting food was no reason at all and should cast around for a better explanation. She ought to reject wanting something to eat as the reason—indeed as *her* reason for leaving the building. And again, a search for a better explanation would involve probing further into the circumstances.

Might we then be looking for subconscious reasons? In a common-sense way of understanding them (and therefore not in the sense that Freud used the term "unconscious") subconscious reasons might be reasons we attribute to an agent that would render her actions intelligible even if they are not the reasons the agent would attribute to herself. We might be inclined to say, in cases like this, that her real reasons for acting were subconscious ones. But, on the common-sense reading, this is tantamount to no more than saying that we have been able to make sense of her actions in her circumstances in a way that she has not. But, *pace* Freud, no notion of anything hidden or inner (like a causal mechanism) need be in play.

There is still another problem about enquiring into what an agent says: her words need to be interpreted. This involves construing her meaning so that it makes sense to us, but now similar conundrums that arise about meaning arise for reasons. Just as it seemed that we needed to make a distinction between reasons that would enable us to make sense of her and her reasons or the reasons for which she acts, we now seem to need to make a distinction between what her words mean and what she means in speaking them. But is the situation not the same for meaning as it is for mental content concepts? One of (the attribution of) meaning's primary uses is to enable us to make sense of what the agent says, just as one of mental concepts' primary uses is to enable us to make sense of what she does. There may sometimes be a point to distinguishing between the meaning of a word and what the speaker means in speaking it, but when there is, we will not have jettisoned the public, social, or "enquirer-centric" pattern of explanation with which we began.[5]

Most of the time an individual is able to answer our questions and remove our puzzlement as to why she performed some particular action or as to what she meant to say without our having to investigate further. But a diagnosis of the source of her authority needs to be handled with care. Her authority is conditioned or constrained by training in a shared discourse and a way of life in which actions and meanings are revealed to be intelligible. I have tried to remind the reader of this by looking at less common cases when the answer she offers fails for some reason to meet the demands of this sociolinguistic practice. There are controls, as Anscombe says, on the truthfulness of someone's proffered reasons and so too are there controls on the truthfulness of his or her proffered meanings.[6]

As Anscombe argues, there may, however, come a point when these controls have been exhausted: there is nothing occurring either before

5. Illustration: When an eleven-year-old child, answering a question on a science exam, responds: "To collect fumes of sulphur, hold a deacon over a flame," I would guess she meant "beaker" by "deacon." When another child taking a history exam writes: "Solomon had three hundred wives and seven hundred porcupines," my guess would be that she really meant "porcupines," though "concubines" was no doubt what she should have said.
6. G. E. M. Anscombe, *Intention* (Oxford: Basil Blackwell, 1957), §25.

or after the action to check on the agent's truthfulness, sincerity, or lack of self-deception in declaring her reasons for acting. Where there is nothing in the circumstances either before or after the action to enable us to pinpoint her reason, then perhaps we reach a point ("after much dispute and fine diagnosis of her genuineness") when only the agent can say why she acts as she does. But this, as Anscombe says, is not because she has access to something—perhaps conceived as an interior movement—that is forever hidden from anyone else, but rather because no one else has any grounds for correcting her. And if she expresses her reasons in advance of acting then most of the time she will have the practical knowledge or ability to go on to act in a way that is explicable by those reasons. But, again agreeing with Anscombe, the credit or condemnation she receives in this regard is that for having performed well or badly—not for being right or mistaken in some judgment. Her knowledge about her reasons for acting in this case is not derived from something that already exists (hidden from others and observable if at all only by her)—it is because she is the author or the creator of her actions.[7]

4. Sometimes, though, we remain puzzled especially when we encounter agents whose behavior is very odd. Suppose, after fleeing the burning building, the firefighter was seen walking into a restaurant and eating a meal. We might have to accept that she left the building because she wanted something to eat. But describing her as wanting something to eat, in this case, is a simple reconstruction of her behavior in the restaurant. The problem is not that we do not understand her behavior in the restaurant, but that we do not understand it in the light of the bigger picture: she was a firefighter who had just fled a burning building. Her reason for leaving the building might have been to get something to eat. But we do not yet have an explanation why a firefighter would leave a burning building to do that. It is not merely

7. Ibid. I develop this further in "Some Constructivist Thoughts About Self-Knowledge," "Self-Knowledge, Normativity, and Construction" and "Conceptual Amorphousness, Reasons, and Causes" (chapters 13, 14 and 16).

that her wanting food does not provide a good or good enough reason for her leaving the burning building: it does not provide a reason at all.

I have spent a fair time examining a particular case in order to get clear about what we actually do when we seek agreement on reasons that explain actions. I have reminded the reader that describing the circumstances is often enough to remove our puzzlement. When we use intentional concepts to explain the actions of an individual, our expressions function, I have argued, not to pick out some first- or second-order physical property of an individual—still less a "psychical" property of hers—but rather as tools that help us keep track of her by carving out aspects of what she says and does that we can, to a greater or lesser extent, understand.

To be sure, there may have been events—hidden from any observer—that preceded the action: introspectible events like the sudden flash of light, the sound of a voice, or a sudden longing. Occasionally, in trying to understand particularly puzzling actions, we will be interested to hear about events like these. ("Why did you murder him?"—"I was intoxicated by the brilliance of the light and the heat of the sun."[8]) More rarely, we might be interested in knowing about hormone changes or electrochemical events in the brain. But this is not in general the case, and when it is, we will have to quit the domain of intentional reason-explanation. Nor is it in general the case that a process of deliberation, reflection, introspection, storytelling or narrative occurs before a fully fledged, language-speaking rational agent acts even though this occasionally happens. So why suppose these processes or something like them are realized in subpersonal, cognitive mechanisms whenever an agent truly acts for reasons? The problem with saying that we do not yet know the real reason, or the reason for which, or the reason that explains the action until we know what conception of the circumstances or what psychological states moved the agent to act is that it gives us a misleading picture. It suggests that there is something missing from, and ultimately, perhaps, unknowable in, ordinary reason-explanations. But this suggestion needs to be resisted.

---

8. See Albert Camus's *L'Étranger* (Paris: Gallimard, 1942).

In the end, we may or may not agree to accept the proffered explanation why the firefighter fled the burning building. Often, we do not agree in our assessment of reasons or in our attributions of intentional concepts. But this is not because we are barred from discovering something that remains hidden from us. It has to do with the very nature of what it is to offer reasons. Wittgenstein remarked that there is in general complete agreement in the judgments of colors made by those who are not color-blind, just as there is in general complete agreement in mathematics (presumably this is because, for controversial cases, color charts and proofs effectively settle the matter). By contrast, Wittgenstein says, there is no such agreement about what another person feels (because no such effective procedures). Nor, he might have added, is there in general such agreement about what another person thinks, wants, judges, or intends; nor about a person's reasons for acting. Is this because our interpretive techniques—those illustrated at some length in this chapter—are too primitive and therefore inadequate? They would be incredibly primitive—tantamount to guesswork—if what it is to feel, think, want, judge, or intend were a matter of the occurrence of an inner mental, physical, or functional event. A better construal is that the techniques we have for establishing agreement about a person's reasons for acting are the best ones for the job, and it is the nature of the practice to elude agreement when there is no longer anything more to say.

After all, mental and other reason concepts may be tools we use to make sense of individuals, but it does not follow that they can always render our subjects understandable. Intentional and other reason concepts do have limited application. The tools are applied but in some cases they are inadequate: just as glue would have little effect on a warped piece of wood. The woman might have left the building in order to get something to eat, but this does not and should not remove our puzzlement as to why a firefighter would leave a burning building to do so.

CHAPTER SEVEN

Reasons as Non-Causal, Context-Placing Explanations

1. It is widely supposed that everyday explanations couched in terms of reasons, motives, intentions, etc. for an agent's actions depend upon law-governed causal relations between states, events, or properties which ordinary mental terms are alleged to pick out or in causal relations between to-be-discovered "realizers" of those supposed states.[1] But this

---

1. This is the root idea of functionalism. Originally, functionalism was proposed as a theory about the meaning of mental terms that are used every day in nontheoretical discourse. David Lewis, "Psychophysical and Theoretical Identifications," *Australasian Journal of Philosophy* 50, 3 (1972): 249–258. Our common sense or everyday concept of pain, on this account, is thought to pick out an (unspecified) inner state of an organism or system that occupies a certain causal role in mediating between other inner mental states, input, and behavioral output. This causal role is to be specified by common-sense platitudes: e.g., that pain is likely to be caused by tissue damage and result in avoidance behavior. Another version suggests that the role is rather to be found by the traditional methods of a priori philosophical analysis. Sydney Shoemaker, *Identity, Cause, and Mind* (Cambridge: Cambridge University Press, 1984). This approach would, like the first, dovetail with the idea that mental concepts' primary domain is common-sense explanation but only if the methods of a priori philosophy are to make explicit what is already implicit in our (common) use of mental concepts. Psychofunctionalists, however, attempt to break the tie with ordinary concepts. They agree with other functionalists that ordinary mental concepts function to pick out a causal role realized by some or other inner

conception of the use of mental terms and of the kind of explanation they serve was disputed by philosophers influenced by Wittgenstein: those who conceived their task to be the untangling of philosophical perplexities thought to arise from inattention to logical or "grammatical" detail. Such philosophers pointed to differences between the employment of mental concepts in everyday reason-explanations and of their (alleged) counterparts in psychological theories—those psychological theories, at least, which resonated to the "comfortingly causal talk characteristic of the hard sciences."[2] They argued that the everyday employment by teachers, lawyers, priests, and doctors of mental concepts explain in a different sense of "explain" from that favored by the hard sciences. Wishing to emphasize the unlikeness between the two senses, these philosophers drew attention to the differences by contrasting reason-explanation with causal explanation and by insisting on important qualifications to the suggestion that reasons, motives, and intentions are causes.

To some commentators, it looked as if these philosophers were taking for granted a hopelessly simplistic and clearly mistaken view both about the concept of causation and about how causation works in nonmental domains. Thus, a number of writers today, though somewhat sympathetic with the writings of Wittgenstein and those he inspired, nonetheless refuse to fight the battle over causation.[3] Although I shall

(physical) state. They also agree that a complete, constitutive account of the second-order or functional states will be given by a story outlining the causal relations between the occupants of such states and their relation to input, output, and other mental states. But they believe not only that the occupants of this role but the role itself are to be discovered by empirical psychology. Georges Rey, *Contemporary Philosophy of Mind—A Contentiously Classical Approach* (Cambridge, MA, and Oxford: Blackwell, 1997), 187; Ned Block, "Troubles with Functionalism," (1978), in *Readings in Philosophy of Psychology,* vol. 1, Block (Cambridge, MA: Harvard University Press, 1980), 268–305.

2. The expression is Jerry Fodor's in *Psychological Explanation: An Introduction to the Philosophy of Psychology* (New York: Random House, 1968), xix.

3. Jennifer Hornsby, "Agency and Actions," in *Agency and Action,* John Hyman and Helen Steward, eds. (Cambridge: Cambridge University Press, 2004); John McDowell, "Might There Be External Reasons?," (1995), *Mind, Value, and Reality* (Cambridge, MA: Harvard University Press, 1998); and Lynne Rudder-Baker, *Explaining Attitudes—A Practical Approach to the Mind* (Cambridge: Cambridge University Press, 1995) are some examples.

not have much to say against these (in many ways kindred) positions, my hunch is that there may be good reasons for resisting the assimilation of importantly different senses of "explain"; such differences, I suspect, tend to be obscured by the appropriation by both camps of the concept of causation.

In this paper, I would like to trace the decline of the ill-understood perspective found in the work of Wittgenstein—paying close attention in particular to the writings of Melden, Ryle, and Anscombe—in order to bring this particular orientation back into view. In section 2, I shall sketch my own understanding of the position, and, in so doing, answer some of the criticism of Davidson and Fodor. In section 3, I shall contrast this view with the instrumentalist and realist ones associated with Daniel Dennett as well as with the behaviorism (mistakenly) attributed to Ryle. I suggest that because on the view I have outlined mental concepts discharge their explanatory role other than by referring to a state, relation, event, or property whose nature is in question, the position I am advocating fails to find its place on the metaphysical map charted by realists and their irrealist opponents. In section 4, I shall offer some hypotheses as to why the construal of reason-explanation as non-causal, *context-placing* explanation has been resisted.

2. Causation, A. I. Melden said, is one of the "snare" words of philosophy.[4] Looking carefully at how this word is used will not allow us to distinguish it from "reason" or even from "explanation": indeed, some use the words "causation" and "explanation" interchangeably.[5] On this use, to deny that the mental is causally efficacious is to deny that the mental is explanatory, and this (correctly) strikes most people as

4. A. I. Melden, *Free Action* (London: Routledge & Kegan Paul, 1961).

5. Compare: "Words like 'explanation,' 'law,' 'rule,' 'principle,' 'why,' 'because,' 'cause,' 'reason,' 'govern,' 'necessitate,' etc. have a range of typically different senses. Mechanism seemed to be a menace [to the possibility of free will] because it was assumed that the use of these terms in mechanical theories is their sole use; that all 'why' questions are answerable in terms of laws of motion." Gilbert Ryle, *The Concept of Mind* (1949) (London and New York: Routledge, 2009), 65. Substitute "physics" and "physical" for "mechanism" and "mechanical" and "physical laws" for "laws of motion" and you have a nice statement of what we might call, with a nod to Ryle, the "bogey of physicalism."

absurd. Nonetheless, there was (and, as we shall see, still is) a widespread attraction to a broadly Humean view about causation—thus it will be worthwhile to bring out the differences between the explanations.[6]

On this broadly Humean understanding, causation is a relation between two logically and temporally distinguishable events. This is sometimes accompanied by the idea that the relation is explanatory insofar as it is subsumable under natural laws or law-like generalizations, which it is the business of the empirical sciences to discover. Let us agree to stipulate for the purposes here that we will understand a causal relation minimally as a relation between two logically and temporally distinguishable events. The position I wish to bring back into focus says that what it is for an action to be in execution of an intention or for it to be explicable by reasons is not a matter of there being a causal relation (in this sense) between intention or reasons and action. If causation is to be thus understood, the pattern in virtue of which a person's intentions, motives, or reasons explain her action is not eo ipso causal.[7]

---

6. I think Anscombe is right in saying the concept of causation is as general as the concept of a factor, so that it is misleading to talk about "the" causal relation. G. E. M. Anscombe, "The Causation of Action," *Human Life, Action, and Ethics: Essays by G. E. M. Anscombe,* Luke Gormally, and Mary Geach, eds. (Exeter: Imprint Academic, 2005), 89–108. Thus, like Melden, I am not particularly interested in defending Hume's account, which has been disputed from every angle. It has, for example, been doubted whether causation is best seen as a relation between events (Hume said between two "objects"); it has been questioned whether necessity is involved, whether a singular causal statement implies a pattern or regularity of any kind, and so forth. But it is roughly this use of "cause" which Davidson had in mind when he argued that reasons are causes and is, as we shall see shortly, accepted by Fodor. It is this use which tempts one to construe verbs like "believes," "thinks," "wants" as picking out mental events or occurrences conceived as hypothetical or theoretical entities. This latter picture is my quarry.

7. This formulation allows us to concede that the concept of intention, for example, may be correctly applied, on some occasion, to designate an event that can legitimately be classed as "mental": a state of a person or an event in her history that can form part of an intuitively plausible causal chain issuing in action (see below for some examples); it denies, *pace* Davidson, that the concept of intention's explanatory value depends on the existence of any such causal relation. Note that these are reasons why the thesis I am defending here is not aptly described as denying either mental causation or the "reality" of the mental: though of course it does deny aspects of mental realism as this is commonly understood. It should be noted, further, that on this view I am recommending the distinction between reasons and

For Melden, the motivation to construe motives, intentions, and reasons as constituents in a causal explanation of action is symptomatic of a misguided attempt to give an account of how an event construed as a mere bodily movement (an arm's rising) can be construed as an action (someone's raising his arm). No further description of the performance in respect of its properties as a bodily movement could possibly disclose that additional feature that makes it an action. This, Melden tells us, is for two reasons. First, the occurrence of a mere bodily happening (say, an event in the brain) does not have the logical force to turn a bodily movement into an action (and this would be so even if events of the one kind enter into lawful relations with events of the same type as the bodily movement to be explained). Second, if we hypothesize the occurrence of something with the right sort of logical force (and call it an intention, a motive, a reason, or a belief-desire pair) then we must, in doing so, presuppose the relation between it and the action that this mental occurrence was invoked to explain. Why? Because in order for motives, intentions, and reasons to be explanatory, they must be motives, intentions, or reasons *for* the action in question. But if motives, intentions, and reasons are introduced in the first place to explain how a bodily movement (the arm's rising) becomes an action (someone's raising his arm) then the specification of the motive or intention cannot simply presuppose the action, on pain of circularity.

> A motive is a motive for some action either performed or considered; hence a motive, far from being a factor which when conjoined with any bodily movement thereby constitutes an action, actually presupposes the very concept of an action itself.[8]

This is the kernel of Melden's argument, but it is not easy to understand and in any case is unlikely to move the contemporary philosopher of

causes is not always a firm one. Following Anscombe we can agree that my hanging up my hat because my host said "Hang up your hat" is one in which the intuitive distinction starts to vanish: it depends on the circumstances whether we would call this a cause or a reason. G. E. M. Anscombe, *Intention* (Oxford: Basil Blackwell, 1957), §15.

8. Melden, *Free Action,* 83.

mind who learned in her first, introductory course on the subject that intentional states enter into both logical relations with other states (in virtue of their content) and causal relations with other states (in virtue of their form). To such a philosopher, Melden's criticism amounts to the denial of a philosophical platitude. So more work is needed, it seems, in order to make the argument clearer.

The gist of the argument I shall develop is this. Correctly to ascribe an intention, motive, or reason in such a way as to display its logico-grammatical relation to action is already to attempt an explanation of the action by putting it into a context that makes it understandable. To suppose that there are events that are designated by the reason- or motive-expression is not only unnecessary; it obscures the way reason-explanation functions.

Davidson claims that it would be a mistake to conclude from the fact that placing the action in a larger pattern explains it, that we now understand the sort of explanation involved. "Cause and effect," he argues, "form the sort of pattern that explains the effect in the sense of 'explain' that we understand as well as any."[9] Davidson challenges the opponents of the causal view to identify what *other* pattern of explanation illustrates the relation between reason and action if they wish to sustain the claim that the pattern is not one of cause and effect. Let us try to meet this challenge. We begin, I maintain, by *assuming* that motives, intentions, and reasons can often be used successfully in explanations of actions and then we ask, when they are successful, how they explain. In many cases, attributions of motives, intentions, and reasons explain a performance by characterizing it as an action of a certain kind. This is already to distinguish an explanation in terms of motives and intentions from a causal explanation, Melden tells us, since a causal explanation suggested by the Humean picture usually takes it for granted that the event to be explained is already fully characterized as the kind of event it is; a causal explanation offers us "an account of how it is that an event whose characteristics are already known is brought to pass."[10]

9. Donald Davidson, "Actions, Reasons, and Causes," (1963), in *Essays on Actions and Events* (Oxford: Clarendon Press, 1980), 10.
10. Melden, *Free Action,* 88.

Of course it is consistent with this, as Melden himself immediately acknowledges, that the effect-event can be described in terms of its cause as an injury to the shoulder, say, might be described as a sunburn. But there must, on this view, be two logically independent (and therefore independently describable) events that enter into the causal relation. This is a condition that Davidson accepts.[11] So, too, does Fodor:

> It is, of course, true that if *X* *is* the cause of *Y*, then there must be *some* description that is true of *X* and that is logically independent of the description "*Y*'s cause," and there must be some description that is true of *Y* and that is logically independent of the description "*X*'s effect."[12]

Fodor adds, however, that this demand would be satisfied if the materially sufficient conditions for having a certain motive could be formulated in neurological terms; indeed, it would be satisfied by the existence of *any* state of affairs that is associated one-to-one with a psychological state by laws, empirical generalizations, or even by accident. Thus, Fodor alleges, the appeal to Humean strictures is too weak for Melden's purposes.

Both Davidson and Fodor seem to interpret Melden's claim that a cause must be "logically distinct from the alleged effect"[13] as dictating the vocabulary that must be used to pick out the supposed mental event which—in order for it to count as a cause at all (it is agreed by everyone here concerned)—must have some logically independent description (whether we know what it is or not). By contrast, I read Melden as calling into question the idea that such a mental event or occurrence must exist. His argument, as I see it, is that the existence of such occurrences is not required for the concepts of intention, motive, and reason, etc. to discharge their explanatory role, thus throwing into

---

11. See Davidson, "Actions, Reasons and Causes," 12, where he suggests a number of candidates for such an event.

12. Jerry Fodor, *Psychological Explanation: An Introduction to the Philosophy of Psychology* (New York: Random House, 1968), 35.

13. Melden, *Free Action,* 52.

question the whole idea that this explanatory role is causal. This, in any case, is the argument I shall develop.

In order to bring to light some of the features of a contrasting, non-causal pattern of explanation, let us consider a simple case first—one removed from the context of reasons, intentions, and motives. A chemistry student who had to leave the class early might find it puzzling why his teacher wrote "cat" on the board. We can imagine his puzzlement relieved when his classmate explains, "Because she was writing "catalyst"—you left the room before she completed the word." This "because" introduces an explanatory context, but it is not the sort of explanation in which one event (logically independent or not) follows another. Intuitively speaking, there is one event (the writing of "catalyst") which has not been understood.[14] The answer serves to recharacterize what happened so that it—as newly described—is no longer puzzling. The chemistry teacher's writing "catalyst" on the board is, I assume for the sake of the example, more understandable than her writing "c," "a," and "t." This is not because we have now made out any mysterious connection between the occurrences of two contingently related events—the writing of "c," "a," and "t," on the one hand and the writing of "catalyst," on the other. For even if these were considered (implausibly) two distinct events, they would not be contingently related: writing the English word "catalyst" entails writing the letters "c," "a," and "t." Nor is there any reason to expect that we may find some other description of the performance of writing the letters that would qualify it as a logically independent event, in such a way that events of this newly described kind enter into a law-like connection with events typed as the writing of the word "catalyst." The performance was puzzling only because it was conceived or described as the writing of the letters "c," "a," and "t" or as the word "cat" instead of the word "catalyst." The teacher's writing "catalyst" on the board is not puzzling, I assume, because it is part of a general pattern of action

14. The student may have construed it as the writing of the word "cat" or as the writing of the sequence of letters "c," "a," and "t." The performance may have been puzzling on either construal.

or behavior that "belongs to" or "is at home in" a chemistry lesson. The "because" in "she wrote the letters 'c,' 'a,' and 't' because she was writing 'catalyst'," then, signals a different pattern of explanation from the causal pattern in which one event follows another. Here we have a clear-cut case of a non-causal, context-placing explanation.[15] This case can be used as a model to develop an elucidation of the explanatory role of our concepts of intention, motive, and reason.

Melden's famous example concerns a man who raises his arm. To the question, "Why did he raise his arm?" the answer "In raising his arm, he intended to signal" serves to recharacterize a performance first described as the driver's raising his arm as an act of signaling. "Raising his arm" and "signaling" are different descriptions, each with different "implication threads" (to borrow an idea from Ryle).[16] Although there may be any number of (muscular, physiological, neuronal) events leading up to and forming part of a causal chain resulting in the arm's rising there is no reason to characterize (or identify) any of these events as the motive, intention, will, or reason to raise the arm. Such a characterization in any case would not permit the redescription of the arm's rising as either the driver's raising his arm or as the act of signaling without adverting to the very background circumstances that I am here

---

15. *If* part of Davidson's challenge here is to say how writing "catalyst" on the board "belongs to" or "is at home in" a chemistry class, then it must be admitted that not much more can be said (except one that issues reminders about the kinds of things one studies in chemistry); at least there is no answer that can be given in more fundamental terms. One of the convictions of the position I am defending is that the ability to see actions as fitting into familiar patterns comes about through training and through a shared form of life and not in general through explicit instruction or through prior theoretical (rule-following) operations. Anscombe's account of an intentional action as one for which a certain sense of the question "Why?" is given application is a sophisticated attempt to describe in more detail what "belonging to" or "being at home in" involves in relating the action to the agent's motives and reasons, but it, too, appeals to our (considered) judgments about what makes sense without attempting to explain this. For a discussion of the process of acculturation that enables us to see actions in new ways, see the final chapters of Melden, *Free Action.* For an argument *against* explaining this ability in terms of prior theoretical operations, see "Playing the Rule-Following Game" (chapter 3).

16. See Gilbert Ryle, "Philosophical Arguments," (1945), in *Collected Papers,* vol. 2 (London: Routledge, 2009).

trying to show may be sufficient for such a redescription, and thus for a non-causal, context-placing explanation.

Wittgenstein thought that we tend to be misled into thinking that there must be such an event—even if hidden from view—because we are focusing on one way that language functions to the exclusion of others. The concepts in question do indeed allow us to speak, for instance, of a person who comes to a decision, forms an intention, or admits that such-and-such reasons for acting are overriding. These uses encourage the thought that having an intention is the product of having made a decision or that having reasons to act results from having considered and accepted certain propositions; thus, intentions and reasons are construed, understandably, as mental events. Now it is true that a full elucidation of the concepts of intention and of reasons and their cognates would have to include these uses.[17] The formation of an intention or the consideration and acceptance of reasons might also figure in a causal explanation of an action.[18] But it would be a mistake to form a

---

17. See Anscombe, *Intention,* for such an elucidation; see "Self-Knowledge, Normativity, and Construction" (chapter 14) for my own attempt.

18. See "Conceptual Amorphousness, Reasons, and Causes" (chapter 16). Anscombe gives an example of such a story. She imagines a case in which she has a long-standing resolution never to grant interviews with members of the media. When someone asks why she refused to see the representative of *Time* magazine, he is told of this long-standing resolution, which "makes her reject such approaches without thinking about the particular case." This explanation, involving the expression "makes her . . ." is causal, says Anscombe, in the sense that it derives the action from a previous state. Anscombe, "The Causation of Action," 95. Or, to borrow an example of Rogers Albritton, my recognition of someone's character, for example, might cause me to break off relations with him. But, warns Anscombe,

> [i]t is one thing to say that a distinct and identifiable state of a human being, namely his having a certain intention, *may* cause various things to happen, even including the doing of what the intention was an intention to do; and quite another to say that for an action to be done in fulfilment of a certain intention (which existed before the action) is *eo ipso* for it to be caused by that prior intention. ("The Causation of Action," 101)

In other words, an event (say) in an agent's history that can legitimately be classed as mental (e.g., his having made the decision, in the light of various factors, that he must do such-and-such) may feed into a causal story of a subsequent action that is performed in execution of that intention. But it should not be inferred from this that what it is to act in execution of that intention is a matter of there being some causal relation between this general picture of the nature of intention or the nature of reasons from these uses alone and thus it would be a mistake to require that every time we ascribe one of these concepts, there must have been a moment when the intention was formed or the reasons considered. Davidson argues for the former in his account of intention (1978);[19] those working in the spirit of Davidson today have argued for the latter.[20]

We have seen that a rather different way of understanding the explanatory power of the response "He intended to signal" is by the placement

---

event and the action. One unhappy consequence of making this inference would be the supposition that there must have been such an event—possibly hidden or nonconscious—even when there is no obvious, conscious candidate. The following closely related idea may be useful in helping to understand this point. It may be that in some particular performance that counts as following a rule, a person consults an expression of that rule and then acts as it mandates. But it should not be inferred from this that what it is to follow a rule is a matter of there being some (overt or hidden) consultation of a rule. I discuss this further in "Playing the Rule-Following Game" and "Real Rules" (chapters 3 and 4).

19. In Davidson's "Actions, Reasons, and Causes," acting for a reason consists merely in having a belief and pro-attitude with the right sort of content, which cause the action. Davidson responds to Melden's challenge to find the mental event that constitutes the reason (understood thus) by declaring:

> Of course there was a mental event; at some point the driver noticed (or thought he noticed) his turn coming up, and that is the moment he signalled. . . . To dignify a driver's awareness that his turn has come by calling it an experience, or even a feeling, is no doubt exaggerated, but whether it deserves a name or not, it had better be the reason why he raises his arm. (12–13)

Thus the idea that the relevant concept's function is to designate some kind of mental event is in place early in Davidson's work. Incidentally, the driver's noticing his turn coming up would (at best) be a reason why he chose that moment to signal.

20. Consider David Velleman:

> In order to have acted autonomously, the agent would need to have been actuated not only by the desire and belief mentioned in the story but also by the story itself, serving as his grasp of what he was doing—or, in other words, as his rationale. He would need, first, to have been inhibited from acting on his desire and belief until he knew what he was up to; and then guided to act on them once he had adopted this story. He would then have acted autonomously because he would have acted for a reason having been actuated in part by a rationale. (David Velleman, *The Possibility of Practical Reason* [Oxford: Oxford University Press, 2000], 28)

Michael Bratman's claim that (full-blown) actions are caused by higher-order reflexive policies is similar to Velleman's. See Michael Bratman, "Two Problems about Human Agency," *Proceedings of the Aristotelian Society* 101, 3 (2001): 309–326.

of the performance first described as the man's raising his arm in the wider circumstances of his driving a car and being about to make a turn. The response will succeed in explaining the man's raising his arm, however, only to the extent that a description that puts it into this context is more understandable than a description that leaves it out. If the one who is puzzled does not understand our driving practices or why anyone who is driving and approaching a turn should signal, then this recharacterization of the action will not satisfy her.

Now Davidson acknowledges that a logico-grammatical relation is in place between the contents of the relevant attitudes (the belief and pro-attitude which, for him, constitute the agent's reason) and the action-type that it recommends. He also holds—what I am here calling into question—that attitude or reason-ascriptions function by designating events (or standing states and triggering events). His (positive) argument for construing this relation as causal is that the logico-grammatical relation exhibited in the content-description is an "anaemic" justificatory one: insufficient for accommodating the case in which the agent has a reason for acting in a certain way, does so, but not for that reason. For example, even though the driver had reason to signal—he was approaching his turn—he may have raised his arm for another reason—say, to wave to his friend. If so, then the recharacterization of his action as a case of signaling will fail to explain the action. Davidson presumably had this sort of case in mind as a counterexample to Melden.[21]

The fact that a context may be imagined, however, in which the redescription fails to explain the action presents no threat to the argument. For it is no part of Melden's job to insist that every context-

---

21. See "Actions, Reasons, and Causes," 11. I suggest in "Why Reasons May Not Be Causes" (Essay 5) that what Davidson is really after is a sort of logico-causal "cement" for what he takes to be a logical gap between reasons and actions. (I have also suggested that his adherence to a Hempelian nomological-deductive model of explanations looms large in the background here.) On the view I am recommending this logical gap must not be closed. Reason does not determine, or provide a sufficient condition for, the action that it explains. Nor does a performance guarantee that a rationalizing, reason-explanation is on offer. The relation between reason and action is more the relation between warrants and moves than the "determinate connection" suggested in the causal, nomological-deductive account.

placing redescription will succeed. Melden need only argue that when such an explanation does succeed (because it enables the one who is puzzled to see the action in a new, sense-making light) it may be the kind of context-placing explanation just described; one that does not depend upon or cannot be understood as requiring the existence of mental events—let alone (in principle describable) logically independent ones—that are alleged to constitute the reason or intention. And, although it is true that the redescription would not succeed unless the requisite motives, intentions, beliefs, and desires could also be ascribed, there is no obligation to construe the deployment of these related concepts as the identification of events or standing states—let alone (in principle describable) logically independent ones. When context-placing explanations such as these are unsuccessful we may need to probe further for a different or more far-reaching context-placing explanation that will succeed or possibly give up the initial expectation that the action can be explained by reasons—but not assume that having reasons, intentions, and motives must be a matter of the instantiation of properties which may be in some sense hidden (e.g., tokenings of conscious or nonconscious mental events or of their alleged realizers).[22]

The problem in assuming that the motive, intention, or reason is (in principle describable as) a logically independent, temporally antecedent, causally efficacious event (perhaps identified with its alleged "onset") is that it misassigns the explanatory function of these concepts. The position commits us to postulating an event, unobservable to others and possibly even to the agent herself that would, if known, provide the sought-after reason-explanation for the agent's action. In such cases, as Ryle insisted, an epistemological puzzle arises how anyone could *ever* know whether a person acts for reasons or what, if she does, her reasons are, since the hypothesis is not even in principle testable. Not only do we not, in everyday situations, have access to these hidden events, but even if we were, say, to monitor the neural activity of someone's brain or access their stream of consciousness, we would never be able to set up the kinds

22. See "Why Reasons May Not Be Causes" and "Reason-Explanation and the Contents of the Mind" (chapters 5 and 6) for extended arguments for these claims.

of correlations that would establish a particular occurrence as an instance of a particular reason without already having a way of deciding whether someone acted for a particular reason in order to make the correlation.

The foregoing considerations suggest that mental concepts such as intention, reason, and motive operate very differently from causal concepts—say, that of a gene. We might say that T. H. Morgan's concept of a gene was the concept of something whose nature was to be discovered, responsible for the transmission of heritable characteristics. The DNA molecule, it was later found, plays that role. But the argument of this paper is that the concepts of intention, reason, etc. are not like this, for they discharge their explanatory role without designating anything: let alone causally efficacious states or events: let alone causally efficacious states or events whose nature awaits discovery.[23]

3. THIS, I THINK, is the correct way to understand the arguments of Wittgenstein, Melden, Anscombe, and Ryle, though it is difficult to know how to place this view within the contemporary debate in the philosophy of mind because it so rarely makes an appearance in today's discussions. A student in philosophy of mind might ask, for example, if this makes the view about intentions realist or irrealist, instrumentalist or behaviorist. In order to facilitate my aim of reintroducing Wittgensteinian territory into the contemporary landscape, it will be worthwhile taking a brief look at the temptation to plot this position with a particular metaphysical compass and suggest a reason why this inclination should be resisted.

We were introduced to instrumentalism in the early work of Daniel Dennett. In "Intentional Systems" he describes the intentional stance by considering a chess-playing computer. In taking the intentional stance toward this machine, one is *instrumentalist* about propositional attitudes insofar as "we find it convenient, explanatory, [and] pragmatically necessary for prediction, to treat it as if it had beliefs and desires

23. For early attempts to argue that mental concept should be construed by analogy with the concept of a gene, see Fodor, *Psychological Explanation* and Hilary Putnam, "Brains and Behaviour," (1965), *Readings in Philosophy of Psychology*, vol. 1, Block, 24–36.

and was rational."[24] A machine for playing chess, however, is not like a man or animal: "its 'rationality' is pinched and artificial."[25]

This was Dennett's position in the 1970s. A decade or so later, his position seemed to change:

> [*A*]*ll there is* to being a true believer is being a system whose behavior is reliably predictable via the intentional strategy, and hence *all there is* to really and truly believing that *p* (for any proposition *p*) is being an intentional system for which *p* occurs as a belief in the best (most predictive) interpretation.[26]

These two characterizations are, on the face of it, inconsistent. On the first, there really is something to being a believer over and above being predictable by the intentional stance, and on the second there is no more to being a believer than being thus predictable. On the first characterization there seems to be an implicit acceptance that mental terms like "believes" pick out underlying, possibly causally efficacious states or events. According to this picture, there is a difference between low-grade computers and people: people really have the underlying states (etc.) to which mental terms purport to refer. Instrumentalism on this construal is like an "error theory" in Mackie's sense. Just as, for Mackie, moral terms purport to pick out moral facts, but do not really (and nonetheless serve their jobs) so do mental terms purport to pick out inner states, but sometimes do not really (and nonetheless serve their jobs). On the second characterization, however, there may be no implied commitment to such underlying states in our use of mental

---

24. The discussion up until now has concerned intentions, reasons, and motives whereas this paragraph introduces beliefs (and propositional attitudes in general). This is not the place to defend or elucidate the idea that the relation between these concepts and the concepts of reason, intention, and action is a (logico-) grammatical one, but see "Reason-Explanation and the Contents of the Mind" (chapter 6) for one such discussion and the last chapters of Melden's *Free Action* for a different discussion of how the concepts of agency, want, and belief are thus connected.

25. Daniel Dennett, *Brainstorms—Philosophical Essays on Mind and Psychology* (Montgomery, VT: Bradford Books, 1978), 8.

26. *The Intentional Stance* (Cambridge, MA: A Bradford Book, MIT Press, 1987), 29.

expressions (on the contrary—this seems to be denied by the locution "*all there is* to being a believer . . ."). If there is no such commitment, this view would not be instrumentalist, since there would be nothing real to contrast with what is supposed to be instrumental.

Is the second position a form of realism then? According to Devitt and Sterelny, Dennett's later view is a form of "philosophical behaviorism."[27] Although usually construed as a type of fictionalism or instrumentalism about the mental, philosophical behaviorism is understood by them, in the context of discussing Dennett, as the realist doctrine that mental terms refer to (real) patterns of behavior. But however appropriate or not this might be as a description of Dennett's later position, it would *not* be a fair description of the view I am attempting to bring back into focus. For on this view, to ascribe an intention, motive, or reason for some particular action may not involve an attempt to refer to anything; the concepts may function rather to explain an action by placing it in a context that renders it less puzzling.[28]

Nor, for similar reasons, would it be a fair to characterize Ryle's view in *The Concept of Mind* as a form of philosophical behaviorism even though he is widely (and misleadingly) credited with introducing us to this doctrine.[29] Ryle's dispositions play the same role as the "sense-making pattern" or the "wider circumstances" play in the view I have just described. Ryle agrees with his interlocutor that when we use mental concepts to describe a performance, we are not merely taking into account "muscular behaviour," because the same muscular behavior in other circumstances could not be so described. A remark by a parrot, for example, could not be described as intelligent or witty. But it does not follow that

27. Michael Devitt and Kim Sterelny, *Language and Reality—An Introduction to Philosophy of Language,* 2nd edn. (Oxford: Blackwell, 1999), 293.

28. Note that to say that the concepts' function is to explain in the way described above is very different from saying that the concepts discharge their role by designating a state, relation, or process which is to be identified by its functional/causal-explanatory role.

29. I argue for this in more detail in "Une Cartographie des Concepts Mentaux," Critical Introduction to Gilbert Ryle's *La Notion d'Esprit* (Paris: Payot, 2005), 7–70 (French version); "Rethinking Ryle: A Critical Introduction to *The Concept of Mind*" in Gilbert Ryle, *The Concept of Mind* (London: Routledge, 2009), ix–lvii (English version); and in "Gilbert Ryle," *Stanford Encyclopedia of Philosophy* (2007) http://plato.stanford.edu/entries/ryle/.

in order to be credited with wit or intelligence the muscular behavior must be accompanied by some mental act. In judging that a particular performance is intelligent, it is true that we look beyond the performance itself: but not into some "hidden counterpart" performance, taking place behind the scenes. We are considering, according to Ryle, the abilities and propensities of which this particular performance was an actualization. "Our inquiry is not into causes (and *a fortiori* not into occult causes), but into capacities, skills, habits, liabilities and bents."[30] For Ryle, many of our "mental-conduct" verbs are correctly applied to a performance or an action because it is the actualization of a disposition.

Talk about dispositions, however, complicates the matter. According to Ryle, the particular mental conduct terms whose logical geography he was attempting to map discharge their explanatory role by helping to situate the agent and her actions within a pattern that can be articulated by an indefinitely long series of hypothetical (and mongrel-categorical) statements about what she could be expected to do, think, feel, etc., given her background (e.g., training) and the present circumstances. This is what he means in this context by noting that mental-conduct verbs are applied to performances that are actualizations of a disposition.

The introduction of dispositions, however, will for others take us right back to the realm of hidden, underlying causes.[31] Quine, for example, was dissatisfied with dispositional statements because they, like general causal statements, depend upon an intuitive or unanalyzed notion of similarity or kind. Dispositional statements, best understood as subjunctive conditionals, are not amenable to paraphrase in the canonical (extensional) language in which Quine held that all serious scientific statements could be expressed.[32]

Quine suggested that when the disposition is of theoretical interest, then a mature science can relinquish this intuitive similarity notion by

30. Gilbert Ryle, *The Concept of Mind,* 33.

31. See Stephen Mumford, *Dispositions* (Oxford: Oxford University Press, 1998) who argues for a particular (functionalist) version of realism about dispositions.

32. Quine's claim that a serious scientific theory must be expressed in an extensional language bodes ill not only for unreconstructed dispositional statements but also, notoriously, for content and meaning in general.

finding the underlying structure that will tell a more straightforward story in a way that renders the dispositional one obsolete: "sometime, whether in terms of proteins or colloids or nerve nets or overt behavior, the relevant branch of science may reach the stage where a similarity notion can be constructed capable of making even the notion of intelligence respectable. And superfluous."[33] Attraction to some of the aspects of Quine's program may be among the reasons scientific realists look deeper than observable patterns of dispositions and search for a common underlying structure in the kinds of things that manifest those patterns. In certain cases, this may be essential: if the dispositional concept is a theoretical concept it arguably needs the discovery of a "realizer" to vindicate it (as the discovery of the DNA molecule presumably was needed to vindicate the concept of gene).

It is precisely this way of construing mental concepts, however, that is under dispute. When Ryle spoke of the "higher-grade" dispositions of people as "multi-track dispositions the exercise of which are indefinitely heterogeneous," and when he used the example of Jane Austen's representation of pride "in the actions, words, thoughts and feelings of her heroine in a thousand different situations,"[34] he was reminding us of the patterns of conduct with which we are already familiar: indeed, he insisted that "the concepts of learning, practice, trying, heeding, pretending, wanting, pondering, arguing, shirking, watching, seeing and being perturbed are not technical concepts."[35] It was not part of his project (any more than Jane Austen's) to speculate about the underlying structure of the "systems" or of people who exercised these dispositions. Nor did he think that such a scientific project could vindicate—let alone in principle replace—the everyday attributions effected by ordinary mental terms.[36]

33. Willard V. O. Quine, "Natural Kinds," (1969), *Naming, Necessity, and Natural Kinds,* Stephen P. Schwartz, ed. (Ithaca, NY: Cornell University Press, 1977), 174.

34. *The Concept of Mind,* 32.

35. Ibid., 319

36. For more on this topic see "Pain, Polio, and Pride: Some Reflections on 'Becausal' Explanations" (chapter 8).

4. One used to hear the complaint made by philosophers impressed by Wittgenstein's teaching that those who tried to treat mental terms as theoretical posits were guilty of changing the subject. This charge was rarely elaborated so it was dismissed on the grounds that it struck the opposition as sheer philistinism or brute prejudice against science. Indeed, the charge *would* sound philistine to those who have already accepted that psychology provides the science behind the ordinary use of mental concepts (together, perhaps, with an attraction to aspects of Quine's program).[37] But precisely these assumptions are denied by Wittgenstein and his followers by their insistence on distinguishing between different kinds of explanation. According to them, to disregard what is sense-making and observable in preference to what is underlying and whose nature awaits discovery is to overlook and thus misinterpret the role mental concepts play in our interpretive practices: to focus on underlying structures would force a change of subject by ignoring the way mental concepts normally discharge their role. The injunction, for example, to accept the observational-dispositional nature of mental terms for everyday use but to insist that their explanatory role depends on how well they interpret states within the system's underlying structure does not make sense on the view I am recommending. This is because the explanatory function of reason concepts *may* be fully discharged by the placement of the action to be explained within the appropriate circumstances or wider context, or, as Anscombe suggests, so that a certain sense of the question "Why?" is given application. For philosophers such as Melden, Anscombe, and Ryle, what it is to describe an action as one performed for such-and-such reasons or with such-and-such intention *may* simply involve an attempt to redescribe what in the context was puzzling with what in the new context is no longer so. Insofar as the causal hypothesis forces us to construe the reason- or intention-ascription as functioning to designate an event, property,

37. See, for example, Rey, *Contemporary Philosophy of Mind.*

state, fact, or condition of a person *the mysterious nature of which is open to investigation,* it misassigns the concepts' explanatory role.[38]

In any case, the complaint of prejudice can be turned around against those who suppose that the role of truly explanatory concepts must be capturable in a language suitable for the aims of mathematics and logic. Quine, for his part, suggests that "amiable" and "reprehensible" are disposition terms that *should* draw on intuitive kinds. Why not suppose that mental conduct terms are like those?[39]

I have argued that it *may* be the familiarity, unsurprisingness, or sense-making aspect of the context, pattern, or circumstances that perform the function of explaining the action, and it is this pattern that is illuminated when the content of the reason, motive, or intention is ascribed. This "sense-making" criterion is closely related to, but sometimes conflicts with, another criterion we use for ascribing intentions, motives, and reasons: namely, that which the subject herself says or would say about her reasons (etc.) for acting.[40]

---

38. Compare:

> How does the philosophical problem about mental processes and states and about behaviourism arise?—The first step is the one that altogether escapes notice. We talk of processes and states and leave their nature undecided. Sometime perhaps we shall know more about them—we think. But that is just what commits us to a particular way of looking at the matter. For we have a definite concept of what it means to learn to know a process better. (The decisive movement in the conjuring trick has been made, and it was the very one that we thought quite innocent.)—And now the analogy which was to make us understand our thoughts falls to pieces. So we have to deny the yet uncomprehended process in the yet unexplored medium. And now it looks as if we had denied mental processes. And naturally we don't want to deny them. (Ludwig Wittgenstein, *Philosophical Investigations,* G. E. M. Anscombe, trans. [Oxford, Blackwell, 1953], §308)

39. The Quinean motivation for reducing dispositions to underlying structures is dubious in any case. Even if they could be accommodated by logical and mathematical methods, these methods themselves would have to rely on intuitive notions of similarity: this is, after all, the lesson of Wittgenstein's discussion of rules. See ibid., §§143–155 and §185ff. For an admirable discussion of this point, see Ilham Dilham, "Universals: Bambrough on Wittgenstein," *Proceedings of the Aristotelian Society,* 79 (1978–79): 35–58.

40. This may involve either a pronouncement on her motives, intentions, or reasons construed as her own way of making sense of her action, or as a (memory) report of

A fuller treatment of this will have to wait for another time; a few brief (and no doubt provocative words) will have to do. Suppose we have recourse to ask Melden's driver what he was doing: was he signaling or waving to a friend? Sometimes his answer will satisfy us; sometimes it will not. But in answering us, he, too, is attempting to place his action within a sense-making context, for according to the view I am attempting bring to light, he will have mastered the relevant concepts—acquired the various skills—in the same way as everyone else. What if his answer does not make sense of his action? We have a choice. We may accept his answer because in asking for his reasons for acting, we may be seeking (on this occasion) his own conception of his behavior, whether or not it satisfies us. When our way of making sense of him conflicts with his understanding of himself, or with his memory reports of what he was thinking at the time, we may reject his answer as inadequate. That is to say, this second ("self-conception") way of understanding the expression "his reason for acting" may be set aside in favor of—what is on this occasion—the other ("sense-making") one. After all, what a person has to say is not always authoritative: as Anscombe reminds us, there are controls on someone's proffered reasons, motives, or intentions.[41]

Nonetheless, there may come a point when these controls have been exhausted: there is nothing occurring either before or after the action to check on an agent's truthfulness, sincerity, or lack of self-deception in declaring her reasons for acting. Where there is nothing in the circumstances either before or after the action to enable us to pinpoint her reason any better than she has been able to do, then perhaps we reach a point, says Anscombe ("after much dispute and fine diagnosis of her genuineness"), when only the agent can say why she acts as she does. But this is not, as the traditional Cartesian would have it, because she has access to an intention, motive, or reason—now conceived as something interior—that is forever hidden from anyone else, but rather because no one else has any grounds for correcting her. What the agent says is,

---

events that may be accompanying or have preceded the action (e.g., a sudden decision or realization), or both.

41. Anscombe, *Intention,* §25.

in such (unusual) circumstances, the only criterion available—the only sense that can be given—to "her reasons for acting."

The arguments of this paper are intended to support the idea that the explanatory role of mental concepts is different from that supposed in contemporary philosophy of mind. The crux of the debate centers not only on whether mental concepts can be assimilated to theoretical terms: I have suggested that a full treatment of the position would involve denying that the predicative expressions in which mental concepts figure must involve the very notion of a reference or an extension that is at the root of the Carnap-Quine-Davidson program. The arguments I put forward here call into question the idea that mental terms purport to function in the general case as referring expressions: i.e., that their primary use or the way by which they discharge their explanatory role is to designate or name an event, state, object, property, or relation.[42] If I am right, then it would seem that a natural way of conceiving the dispute between mental realists and irrealists is based upon a category mistake from the outset.

---

42. See Ryle, *Collected Papers,* vol. 2, for his explorations on this theme. A number of moves made in metaphysics, epistemology, and in philosophy of language and mind in the 1960s and '70s—ones that are presupposed in most of the work in these fields today—would be thrown into question as well if this idea is correct. If they are not referring expressions, then the construal of them by scientific realists on the analogy with natural kind terms such as "gold" or "tiger"—whose essence is a matter for science to discover—is a nonstarter. This is, for example, the treatment that Putnam, "Brains and Behaviour," suggests for the concept of pain. The diagnosis would also cast doubt on David Armstrong's characterization of conceptual elucidation and ontology as the investigation of second- and first-order questions, respectively. See *The Nature of Mind* (Queensland: University of Queensland Press, 1980), 16–17. So too, of course, would it help to define the real trouble with functionalism. The functionalists (and conceptual-role theorists) were right to focus on the importance of function, role, or use. But the explanatory role is played, I have argued, by the way the concept is wielded in redescribing the behavior of a system. It is not played by referring (however obliquely) to the system's internal (first- or second-order) states. And finally the view that "criteriological" investigations modeled after Wittgenstein's are about justification *and therefore about epistemology and not metaphysics* would founder as well. See the preface to Shoemaker, *Identity, Cause, and Mind,* for a biographical account of how his acceptance of this distinction led him to abandon criteriological investigations for causal accounts.

CHAPTER EIGHT

Pain, Polio, and Pride: Some Reflections on "Becausal" Explanations

Both Wittgenstein and Ryle pointed to differences in the types of explanations provided by the ordinary employment of mental terms, on the one hand, and the style of causal explanation characteristic of the sciences, on the other. Their arguments, however, are ill understood. Their position does not, for example, find its place on a metaphysical map that charts the territory disputed by mental realists and their irrealist opponents. Their view, as I see it, is that the normative dimension along which our various doings, sayings, thinkings, and perceivings can be assessed should be understood, not as normative properties of "mental states"—the nature and explanatory role of which remain mysterious—but as embedded in concepts that are themselves sophistications upon ever-ascending levels of discourse, or sociolinguistic practices.

Without denying that within the domain of the mental we can find items legitimately classified as states or events which may form part of a plausible causal story in which they culminate in, and partly explain, some particular performance, I have, in other chapters in this volume, suggested resisting a commonly held stronger position: that in order for mental predicates to fulfill their explanatory role, the "inner" (content-bearing) states, events, or processes that are considered to be the standard referents of mental predicates must enter into causal relations with other such states, environmental stimulus, and behavior. This is because I recommend rejecting the underlying assumption that mental terms function, in general, to designate something in the first place, a fortiori something with (what is widely thought to be) the appropriate ontological status to play the causal-explanatory role envisaged: namely, an event, say, which has—whether or not we happen to know it—a description that is logically independent of the effect-event that it explains. If this—much or little—is taken for granted in one's view of causation and causal explanation, then the explanatory power of reasons, intention, motive, and cognate concepts, is not causal.

The contrasting view is best approached, I have argued, by investigating how we strive to reach agreement about one another's reasons, beliefs, desires, intentions, and motives. One natural thought is that we ascribe reasons to one another, as we do intentions, beliefs, hopes, wants, etc., as part of our attempt to make sense—and this includes giving appropriate descriptions—of what the other says and does. By considering intentions and motives we may redescribe a killing as a murder, a redirected gaze a snub, a breaking of an heirloom an act of revenge. We might think of an investigation of another's reasons or intentions as one in which we are equipped with a special camera that is focused on the performance to be explained but which we do not yet know how to describe. Trying to make sense of it may lead us to a (perfectly familiar) pattern in which reasons, motives, intentions, etc. are relevant. This requires understanding the circumstances in which the performance takes place, sometimes reaching back into history (in the case of revenge) and forward into the future (in the case of a slow, deliberate murder). Very rarely, however, is this sense-making pattern gleaned from hypothesizing the occurrence of thoughts running through the mind (in the sense that a tune might play in the head) and never, at least in ordinary circumstances, of neuronal activity in a particular region of the brain. Of course, the agent's own explanation is important—not, in general because she has access to inner causes, but because her actions, as well as her explanations, are conditioned by membership in a shared practice and discourse in which her doings and sayings are revealed to be intelligible. Even if we favor our own explanation over hers—perhaps we attribute to her an intention or motive that she does not avow—we learn something about her when she lets us know the description she favors.

Taken seriously sixty years ago, this position is today invisible to those doing cutting-edge work at the intersection of philosophy of mind, language, and moral psychology—especially as these subjects join hands with empirical research that is conducted along lines championed in the cognitive sciences. Since the 1960s philosophers of mind have accepted the "causal efficacy" of the mental and with it a version of mental realism that acknowledges the existence of event-tokens that, though falling under mental kinds or instantiating mental properties, also fall under physical kinds. These token-events are thought to enter into external causal relations, or relations that exist in the world, as it were, independently of any particular descriptive apparatus with which we pick them out. Acceptance of this doctrine is at the heart of both weak and strong forms of physicalism, and incorporates versions of functionalism, as well as its descendants in causal-role, computational, and broadly representational theories of mind.

Hilary Putnam was instrumental in preparing the ground for this doctrine and its offshoots when, in an early influential article, he dislodged the then prevalent view—one I wish to bring back into focus—that mental explanation of action is, in a sense ill understood, conceptual. He suggested we understand mental concepts such as pain on analogy with concepts such as Polio or Multiple Sclerosis. According to his view, though behavior plays an essential role in enabling interpreters (obliquely) to refer to hidden causes, these (assumed to be inner) causal events are only contingently connected to their behavioral effects. This leaves open the possibility in principle that these mental event-causes have an existence, essence, or nature that is ripe for further, empirical investigation that need not, indeed should not, be constrained by the ordinary, day-to-day deployment of mental predicates. It may be that mental concepts allow us to home in on, or pick out mental events that causally explain their behavioral effects, but once they are picked out in this way, it would be up to our best science to help us understand their true natures. Mental concepts fix the reference of mental events or states, whose nature it is for the psychologist or neurologist to discover.

In this chapter, I begin an exploration of some of the philosophical assumptions about language, description, and explanation that lurk in the background of the views in dispute. I suggest, contra Putnam, that medical syndromes such as AIDS, CFS, and Polio form a subset of context-placing explanations which do not have the reference-fixing role he imagines. Saying this, of course, does not obviate the need to identify one or more viral (or other) origins for these syndromes, which in turn may enable us to discover through experiment and experience an underlying or microstructural explanation of the diseases' effects. There is no doubt this increases dramatically our understanding of the syndrome in the sense that it puts us in a better position to manipulate (prevent or cure) it. But classifying something as a syndrome or disease does not serve as a mere placeholder for that of a microstructural explanation: it plays a sense-making or explanatory role of its own. Diseases, understood as syndromes, therefore have something in common with explanations invoking character traits such as pride. I reconsider what Gilbert Ryle meant by his claim that mental concepts are dispositional, and argue that, in spite of the fashion to construe dispositions realistically, Ryle was right to insist that to construe the conceptual story as a form of causal explanation would amount to a type-error or category mistake.

1. In "Brains and Behavior," Hilary Putnam claims that he is out to bury any remnant of logical or philosophical behaviorism.[1] Although he takes his target to be the claim that there are "analytic entailments" between mind-statements and behavior-statements, his arguments, if successful, seem to threaten the thesis that I shall defend: that natural or characteristic expressions in behavior have privileged seats, as it were, at the center of the conceptual domain of the mental. That such natural expressions are at the center of the mental manifold

1. Hilary Putnam, "Brains and Behavior," (1965), in *Readings in Philosophy of Psychology,* vol. 1, Ned Block, ed. (Cambridge, MA: Harvard University Press, 1980), 24–36.

is consistent with the thought that we may ascribe mental predicates in their absence. But when it comes to backing or challenging third-person and even some first-person ascriptions of mental predicates, these expressions or the absence of them—present, past, and future—will remain a major, if not overriding, consideration.

Putnam suggests, when considering the *intension* of pain, that it is a cluster-concept in the sense that "the application of the word 'pain' is controlled by a whole cluster of criteria, *all of which can be regarded as synthetic*."[2] In a footnote he clarifies that he means "not only that each criterion can be regarded as synthetic, but also that the cluster is *collectively* synthetic, in the sense that we are free in certain cases to say (for reason of inductive simplicity and theoretical economy) that the term applies although the whole cluster is missing." This is compatible, he avers, with saying that the cluster serves to fix the meaning of the word. Later in the article Putnam produces a thought-experiment intended to show that the concept of pain might correctly be applied to an individual even if the whole cluster of criteria by which we normally justify or defend the application of the concept *never* obtained and *never will* obtain not only for the individual herself but for her whole community (or race). This is because, again, the cluster of criteria that we are presently prepared to use to justify or correct applications of the concept may serve to fix the reference or pick out the state of pain. But the nature of this state is (logically) independent of the ways we have for referring to it.

Putnam is working with a conception of *concepts* which will be teased out and challenged in more detail later. On the view I will recommend concepts are abstractions from extant linguistic practices; there is thus a degree of fluidity or amorphousness to their contours. A philosopher is free to recommend or speculate about the direction these practices might or should take and, in doing so, suggest relevant extensions or revisions to the concepts. If practices change so that the concept of pain has no use then perhaps it will be consigned to history books, to literature, or forgotten.

2. Ibid., 25.

Putnam's suggestion is not the innocuous one that the concept may no longer have the same role in the lives of creatures who have evolved or whose nature is radically different. His suggestion is that the concept functions now in a way that respects the contingencies of his thought experiment: viz, to fix the reference of what may turn out to be (say) a neurophysiological state of the organism. But Putnam cannot propose a theory about the nature of mental states—the alleged designata of mental terms—unless it respects the function of these terms in the practices in which they are employed. (This way of wording it is consistent with the idea that today's practices can evolve or change—indeed, in the light of empirical investigation.) Although I may be prepared to agree with Putnam that there are no analytic entailments between sentences containing mental and behavioral terms I shall argue that he goes too far when he suggests that the logical tie between mental terms and behavior could be completely severed without making a radical break in our practice of using these terms, for it is this connection that gives mental concepts the explanatory power they have: in particular, the concept of pain's role in helping us classify, and thus make intelligible, moaning, writhing, relief-seeking, and other "characteristic" suffering behavior.

Putnam's argument begins, by analogy, with the concept of disease and, in particular, with the case of Polio. According to Putnam, Polio is a disease that was picked out or identified in the first instance by its symptoms or effects. Later, it was discovered that a virus was responsible for the condition. Putnam claims the discovery of a viral origin for Polio led doctors to refine the application of the concept by stipulating that unless the virus is present in an individual, she does not have Polio. The concept of Polio, he says, is the concept of a disease, which is normally but not always responsible for certain symptoms, which were later discovered to have a viral cause. According to this understanding of the concept, we can say that a person has Polio because she has the Polio virus even though she does not have any of the symptoms. We are, accordingly, prohibited from saying that a person who has any or all of the symptoms has Polio, unless she (also) has the virus.

Putnam suggests that the situation is similar with the concept of pain. Just as Polio is the concept of a disease that is normally (but not necessarily) associated with certain symptoms, so too, might the concept of pain be one of a yet-to-be discovered state that is normally (but not necessarily) accompanied by certain behavioral dispositions. Just as a person who has the symptoms of Polio without the virus fails to count as having Polio, so might the person who exhibits pain symptoms not be in pain. Correspondingly, just as a person who has the virus has Polio whether or not she has the symptoms, so too might she be in pain even if she does not, or indeed, even if she is not even disposed—ever—to manifest pain behavior. Putnam treats us to some science-fiction thought experiments about Super-Spartans and X-Worlders that are meant to persuade us that the concept of pain leaves open the possibilities he describes.

2. PUTNAM CONSIDERS the obvious riposte to the science-fiction cases, which is that the word "pain," having lost all connection to suffering behavior, has changed its meaning in such a way that we are no longer operating with the same concept.

Similarly, explains Putnam, one might challenge the analogical argument by claiming that though "Polio" used to mean the presence of particular responses, when doctors made the decision to accept new criteria for its application, they changed the meaning of the word. But Putnam rejects the suggestion that the meaning has changed on the grounds that it would lead to counterintuitive results. For a doctor who said that he believed Polio was caused by a virus before 1908, when Karl Landsteiner discovered its viral origin, would be wrong, not right, since the word "Polio" as it was used then was not always caused by a virus. Doctors who said of a case presented to them that though exhibiting all the textbook symptoms it was nonetheless not one of Polio would have been contradicting themselves, even if we would now say that they had been right after all. So Putnam rejects the idea that the word "Polio" underwent a change of meaning when a new virus origin was discovered.

This is a crucial step in the argument, and will repay close attention, because it masks a certain view about meaning and what this entails about concepts and conceptual change that should, I believe, be challenged. Before doing so, however, let us note that far from being a natural kind concept, which is supposed to fix the reference of something "in the world," independently of how language is used, "Polio" undergoes various shifts, from the name of a disease which, though not a logical construction out of its symptoms, has some clear meaning-relation to them, to the name of the viral origin. This may be more difficult to see, perhaps, because what became known as "the Polio virus" (with three separate antigens) was named after the disease. Furthermore, contrary to what is commonly believed, the virus did *not* typically result in paralysis. Those infected without the accompanying symptoms were people who had what was called the "minor illness." To prevent misunderstanding when the context does not reveal it, clarification is needed if we want to talk of someone who has been infected by the virus but does not have the full-fledged disease.

This is worth pressing, not because I share (early) Putnam's faith in the analytic-synthetic distinction—it should be clear in what follows that I do not—but because the failure of finding analytic entailments (if this amounts to context-independent necessary and sufficient conditions) between a mental predicate and its characteristic expression should not blind us to the idea that bringing behavior under mental concepts—i.e., describing it—is already to begin the job of explaining it, for it puts it in a context that helps to make it intelligible. This is true even though a concept names no corresponding reality, e.g., an object—abstract or real—property, state, or event. Although Putnam is right in noticing the lack of analytic entailments between mental predicates and their characteristic expressions (if we are to understand these as context-independent necessary and/or sufficient conditions), he goes too far in supposing that mental predicates function by referring to something in the world whose nature we know nothing about, for this makes it sound as if we may find out at a future time that we knew nothing about the nature of pain—including, presumably, that it had anything to do with suffering, relief-seeking, and other characteristic behavior. But this is nonsense.

3. It is a standard argument against logical behaviorism that analytic entailments do not provide explanations: this is why it is thought that we need to break out of a circle of meaning (or "rational" relations) in order to find a causal factor that accounts for the occurrence of the phenomenon we wish to understand. This idea is echoed not only in Davidson's arguments that reasons, which reveal actions as rational, also have to function as causes in order fully to explain the action; it can also be found in the rejection of dispositions as explanatory unless we identify, or at least presume to exist, a categorical property or properties, causally responsible for the manifestation of the disposition. The intuition recurs in the idea that diseases are not reportable (or insurable) until an underlying physical pathology has been discovered (presumably to rule out psychosomatic illness). Again, however, the argument from the rejection of analytic entailments to the rejection of conceptual explanation goes too quickly, for everything depends on what we will count as an explanation in the circumstances.

"Why is she unable to get out of bed?" or "Why is she refusing to ask for help?" can be answered by a "becausal" explanation—"Because she has CFS"; "Because she is proud"—which serves to locate the phenomenon in a context in which the similarities and differences it bears to other phenomena can be brought to light. It is simply not true, for example, that we have no further understanding of a person's fatigue if we know only that she has Chronic Fatigue Syndrome. "Because she has CFS" in answer to "Why is she tired all the time?" marks a certain place on a manifold—one which, as we shall see, is not isolated from experience—from which we can form rational expectations. Although we do not yet know its viral (or other) origins, we know enough about the disease (from experience) not to be surprised, for example, if the patient is not refreshed after ten or twelve hours sleep, or if she has a relapse after exercise. In the same way, incidentally, the response "Because it contains a sleep-inducing chemical" (a "dormitive virtue") in response to "Why did that pill put her to sleep?" gives us reason to infer, with varying degrees of confidence, that the pill she took was not ammonia, antacid, or aspirin. Answering a "why" question by mentioning a general disposition—even one that, like AIDS, CFS, or Polio manifests a patent "logical" tie to its symptoms insofar as it "merely" redescribes them—puts us in a position to form certain expectations and rule out others.

None of this is to deny that there are underlying microphysical properties which are implicated in the manifestation of dispositions such as solubility or Polio. (Ryle, who is widely known as an "anti-realist" about dispositions, certainly does not deny it.) The concept of solubility, as well as pain and other sensations, will become enriched as we make discoveries that require further experience and experiment. But such discoveries are compatible, and not in competition, with those showing that the phenomenon satisfies a general disposition. Such discoveries may even take their own place on the conceptual manifold: they may play new roles in the backing and support we give for the application of the concept (this is what is meant when it is said that such discoveries may give us new criteria). This, perhaps, is not unusual in the case of diseases, for discovering the viral or other origins will give us the detail we need to manipulate and control them. But this does not mean that adverting to a syndrome which redescribes the symptoms, without knowledge of viral or bacteriological detail, for example, has no explanatory value.

Consider the similarities and differences with character-trait explanations. When Ryle studied Jane Austen's depiction of Elizabeth Bennett's "actions, words, thoughts and feelings in a thousand different situations" he was entitled to describe Elizabeth's pride as the sort which "combines a dangerous cocksureness in her assessments of people with a proper sense of her own worth." By contrast:

> Darcy is, to start with, haughty and snobbish, a true nephew of Lady Catherine de B[o]urgh. His early love for Elizabeth is vitiated by condescension. He reforms into a man with pride of the right sort. He is proud to be able to help Elizabeth and her socially embarrassing family. He now knows what is due from him as well as what is due to him.[3]

There are no analytic entailments between pride and its manifestations, Ryle tells us, since there is no standard type of action or reaction such that Jane Austen could say "My heroine's kind of pride was just the ten-

3. Gilbert Ryle, *Collected Papers,* vol. 1 (1971) (London: Routledge, 2009), 289.

dency to do this, whenever a situation of that sort arose."[4] More importantly, we should add on his behalf, there are no analytic entailments since the concept of pride is not on the same level of discourse as the actions, words, thoughts, and feelings that manifest it or that satisfy its ascription. The concept of pride "collects" its indefinitely many manifestations; these manifestations are "satisfied" by it, for the concept of pride is a tool, forged by our ancestors, with which we categorize ways of acting, thinking, and feeling. With this concept we put the actions, thoughts, and feelings into a context which makes them more understandable: by categorizing them thus we are entitled to form certain expectations and rule out others. In showing us how her characters fall at various points on the manifold of the concept of pride and comparing and contrasting the various manifestations as an expert would appreciate fine teas and fine wines, Austen, in *Pride and Prejudice,* undertakes a conceptual-cartographical study of it, and her readers' understanding of pride and of its role in human nature is enhanced as a result of their engagement with this novel.

As with the concept of pain, the manifestation remains the grounds or support we give for the application of the disposition-concept. This is the sense in which there is an internal, yet nonentailment, relation between pain and pain-behavior, Polio and its symptoms, and pride and its appearances. Whether or not a new discovery will count as among, or initiate change in, the justifications, backings, or support we give for the application of the concept will depend on whether it helps us mark the affinities and set aside others that for practical purposes we wish to indicate or ignore. Polio, as the name of a disease, is relatively specific though it has several viral strains. CFS, sufferers hope, will cease to be used as a diagnosis when a specific viral or bacteriological cause is found. Pain is, like pride, a concept which is highly generic: in using it we categorize indefinitely many manifestations while ignoring, among other things, the underlying structure of the organism whose

4. Gilbert Ryle, *The Concept of Mind* (1949) (London and New York: Routledge, 2009), 44.

behavior satisfies it. Solubility is also highly generic, and whether it will be replaced by structural "causal" stories, as Quine hoped, will depend upon whether the abstractness and hence the flexibility of the notion of *the soluble* has a use in our practices. But Pride is an example of a disposition-term for which it seems odd to suggest, let alone insist, that an underlying, inductively discovered and possibly law-subsumed relation between events must be found in order to authenticate its credentials as an explanatory mental predicate.

Mental concepts that have their home in reason-explanation are of the same genre. Mental terms do not describe behavioral patterns: they "collect" them. Performances of certain descriptions count as satisfying these concepts: they are what we, observers, use in the justifications and backings of our employments of them. It is in this sense that behavior, action, and (other) mental concepts are logically connected. "Because he is moaning and groaning" provides a warrant for my claim that he is in pain, and this warrant or inference-ticket is context-sensitive; it is not a circumstance-independent premise from which I deduce that he is in pain. Similarly "Because he thinks there will be a downpour" provides a context-sensitive warrant to expect, with varying degrees of confidence, that he will look for his umbrella before leaving the house. It is not a circumstance-independent premise from which the claim that he will look for his umbrella follows.

4. THE THESIS that behavior has a central place in the conceptual manifold of the mental is not to be confused with the view that Wisdom caricatures as "the idea that the relation between *analysandum* and *analysans* in a philosophical analysis is a piece of string stretched between ice-coloured objects that never melt."[5]

According to the traditional view associated with early Plato and with Frege, a concept is the meaning qua object for which a word stands, and a proposition is the meaning qua complex object for which a sentence stands. Concepts are, on this picture, immutable abstract

5. John Wisdom, *Proof and Explanation: The Virginia Lectures,* (1957), Stephen F. Barker, ed. (Lanham, MD: University Press of America, 1991), 65–66.

objects which exist independently of our linguistic practices; indeed, our use of expressions is thought to be guided and explained by our (prior) cognitive grasp of the relevant concepts. An analysis of the concept should provide us with definitions, translations, or other rules for determining how the relevant expressions are to be used, including the phenomena or kinds of phenomena they are supposed to signify. Since the objects alleged to be named by the abstract concept-nouns are independent, not only of our judgments, but of our practices, what we cotton onto when we grasp a concept is (if they determine correct use) logical entailments or necessary and/or sufficient conditions, which (if they are immutable and independent) float free from any particular context in which they are applied.

On Ryle's view, the germ of which he credits to Wittgenstein's *Tractatus* but develops independently of Wittgenstein's later work, the traditional notion of concepts gets things back to front: they are not immutable, determinate, or independent of our actual (contingent and fluid) linguistic practices. As I understand the view, concepts are discernable only through the linguistic practices from which they are abstracted. A particular application of the concept (which takes a certain place on the conceptual manifold) is linked to others not by some one thing that is common to the situations or objects to which it is applied, but by the "overlapping" and "crisscrossing" justifications or backings that we (those who speak the language and engage in the practices) offer and accept when the application of the concept is taught or challenged. Because all if not most of our expressions have indefinitely many "inflections of meaning," the "logical threads" of a statement embodying the concept—i.e., its implications, its compatibility or incompatibility with different affiliated statements, its being evidence for or against different corollaries, the tests required for truth and falsehood, etc.—will change with the particular inflection that it assumes in the circumstances of its employment. What satisfies the concept is thus fluid in one sense. New discoveries and decisions to extend or alter the contours of the concept will usher in new backings and support for the claims. Thus what satisfies the concept is fluid in a second sense as well.

In contrasting this view with the traditional one, in which cognitive grasp of a concept (on this picture, an abstract object signified by an abstract noun) is explanatorily prior to the ability to wield it, we might say instead—to borrow a metaphor from Wittgenstein—that to engage in conceptual investigation (to mark an expression's place within the conceptual manifold and tease out its logical threads) serves rather to identify "the post at which it is stationed." In contrast to the traditional view, to inform someone of the post at which an expression is stationed allows her to trade with it only if she already *knows how* to engage in the permissible transactions. Or to adopt one of Ryle's metaphors, the afternoon task which involves talk about concepts is useful only to those who know how to engage in the morning task—those who know how to operate with these concepts. The ability to wield the concept of time, for example, by using tensed expressions, clocking race cars, and making a commitment not to be late, is a precondition and not a result of a cognitive grasp of the nature of Time.

Putnam's reasoning in this early article seems to be that because there are no analytic entailments between the mental domain and the behavioral then we must reject the traditional view and turn to science. But note the resemblance between the traditional view and the one he champions. On the latter, rather than viewing the referent of abstract nouns as language-independent, stable (abstract) objects, they are to be construed instead as natural kind terms that refer to (language-independent, stable) states, processes, or other type of ontological entity. The suggestion that the meaning of "Polio" would have changed had doctors accepted new criteria for its existence (a viral origin) was rejected by Putnam on the grounds that it was absurd to suggest that before and after the discovery doctors were talking about—referring to—different things. But concept-nouns are not names: either of abstract objects as conceived on the traditional view, nor of "real" properties, states, or processes: either language-dependent or language-independent ones.

Perhaps the implicit reasoning is that if analytic entailments do not exist, then we must reject the thought that our quest involves relations of meaning: what is available through reflection alone. So it must involve matters of fact: what is only available through experience and experiment. As we have seen, however, this is a false dilemma since it seems to be part of the traditional view of concepts with its questionable bifurcation of the "real" world on the one hand and our conceptual resources on the other. On the view I am recommending, investigations of the kind Austen embarks upon in *Pride and Prejudice* for the notion of pride and empirical discoveries about microstructural properties related to Polio or pain may serve to alter the contours of a concept. But as long as substances dissolve in liquid and human frailty and universal suffering continue, then whether or not the application of the concepts of solubility, pride, and pain will change in the light of new evidence will depend upon whether this new evidence strengthens their power—whether, as Wittgenstein says, the discovery throws new light on the old by connecting them with the new, for "the new meaning must be such that we who have had a certain training find it useful in certain ways."[6]

It is, finally, no coincidence that Putnam's stock example of a mental predicate is "pain." For if any mental concept is thought to refer to or describe something the nature of which we can learn about by empirical investigation it would be a sensation-concept. On the one hand, because sensations have a characteristic phenomenology they do not seem to be like dispositional states such as belief or understanding. On the other, if "pain" and other sensation expressions do not name dispositions then, thinking they must name something, we seem forced to conclude they name phenomenological, occurrent states instead. Once these moves have been made, it is difficult to resist the idea that their nature can be uncovered by empirical investigation. There is, of course, a physiology of pain and hunger (in a way there is arguably not for pride or humor), just as there is a phenomenology of both (in a way there is arguably not for belief or knowledge). But the behavioral-dispositional aspects are at the center of the conceptual manifold of pain and hunger, just as they are for belief, desire, understanding, perception, intention, and so on. It is in virtue of this place on the manifold that mental concepts make these behaviors and actions intelligible—by locating them such that their affinities and differences with other phenomena are revealed.

---

6. Ludwig Wittgenstein, *Lectures on the Foundations of Mathematics, Cambridge 1939*, Cora Diamond, ed. (Chicago: University of Chicago Press, 1975), 66.

PART THREE

Philosophical Elucidation and Cognitive Science

CHAPTER NINE

How to Resist Mental Representations

1. Tim Crane's *The Mechanical Mind* is a very readable, introductory book that weaves together various threads of contemporary discussions on mental representation to present a systematic and up-to-date defense of the view that mental states are physically realized, representational states that causally interact to produce other mental states and behavior in accordance with natural, irreducible laws of psychology.[1] Since the early chapters are carefully written in a style that does not presuppose a philosophical background, the later, more difficult discussions of the arguments for and against various ways of naturalizing content should still be relatively accessible to the general reader who starts from the beginning. The book will thus serve as an introduction to the method of, as well as to one of the most important and hotly debated topics in, contemporary analytic philosophy. It will certainly serve as a useful course text for introducing undergraduates or postgraduates to the problem of mental representation and its place in contemporary philosophy of mind.

1. Tim Crane, *The Mechanical Mind—A Philosophical Introduction to Minds, Machines, and Mental Representation* (London: Penguin Books, 1995).

The view discussed here is not original, nor does it claim to be. The merit of this book is that it attempts to mount as strong a case as possible (keeping within the author's aim to attract a nonspecialist audience) by canvassing a variety of familiar sources. The importance of the view cannot be ignored. The belief in physically realized, causally efficacious, content-bearing states has grown steadily and widely in the last twenty-five years, becoming the current orthodoxy in American philosophy of mind and psychology, and gaining wide acceptance in Britain. It has encouraged research programs in the "cognitive" branches of anthropology, biology, psychology, linguistics, and computer science, and has spawned the ubiquitous attempts of the last fifteen years of philosophy of mind to "naturalize content." Since Crane's aim in writing this book is to make this view accessible to the general reader and since he succeeds in delivering one of its simplest and clearest defenses, I think it will be worthwhile to describe the book in detail, and to remind the reader of a strategy for resisting its central claims.

2. THE BOOK begins by looking at the problem of representation in general. What it is for pictures, for example, to represent objects or states of affairs cannot be a matter of mere resemblance: even if resemblance were necessary it does not seem sufficient, since, say, Michelangelo's *Last Judgment* itself does not determine whether Christ is passing judgment or extending his hand in a greeting. Interpretation seems to be needed for pictorial representation, as it seems to be when we ask what it is for words to represent objects or sentences to represent states of affairs. Interpretation is "something the mind bestows on words," and thoughts are what make interpretation possible. So, since thoughts (and other mental states) are representational, too, it would seem that mental representation is more basic than other forms. Since the representational or intentional content of a mental state cannot itself be the result of interpretation (on pain of regress), it must be an example of intrinsic (as opposed to derived) intentionality. A propositional attitude is cashed out as a content-bearing state that may not be conscious. Thus the agenda is set for the rest of the book. A prima facie case has been made for the existence of mental representations, as content-bearing, possibly nonconscious states, which are necessary for an account of representation in general and hence whose own intentional or repr tational properties must be underived. The task now is to understand more about mental representation and how it functions.

The mind–body problem which would have forced us to choose between dualism and materialism (physicalism) is dismissed as irrelevant on the grounds that deciding whether or not the mind is fundamentally different from the physical does not answer the question how one bit of substance (material or immaterial) can represent another. In taking us through some of the other traditional problems in philosophy of mind, Crane uses the discussion of analytical behaviorism to introduce the idea that common-sense psychology is a theory consisting of principles, knowledge of which enables us to explain the behavior (and other mental states) of others.

The behaviorist idea that there is no internal state of the individual that accounts for the disposition to behave under certain circumstances and causes the relevant actions is rejected on the familiar grounds that such a view thwarts the possibility of a (scientific) explanation of the action. Davidson's argument that reasons are causes is marshaled to lend further support to the idea that the mental states posited by common-sense psychology are causally efficacious and Churchland's eliminativist arguments (which depend on the idea that common-sense psychology is a theory) are rejected. We are left with the idea that we find out about other minds by exploiting a theory of mind; we can find out about the nature of the mind by investigating the theory of mind employed. Crane hints that a plausible rendering of common-sense psychology is that it posits causally efficacious content-bearing states that enter into computational relations. It is the task of his next few chapters to discuss in detail what "computational" means in this context.

Before examining the next chapters, it would be worth noting how one might resist even some of these preliminary steps. One may agree that a recalcitrant interpretation might always defeat any candidate (set of) sufficient condition(s) for a sign to represent something without agreeing that interpretation itself is a necessary part of such an account. Thus, the move that forced us to think of mental representations as a necessary component in an account of, and hence more basic than, public representations would be blocked. One may agree, in addition, that we ascribe propositional attitudes in order to make sense of one another: for example, we ascribe them to categorize (rationalize) certain patterns in a person's linguistic and nonlinguistic behavior, taking into account his relations to the social and physical environment. Note that propositional attitudes are often legitimately ascribable even on the assumption that there is no particular interior process occurring (or any particular interior state which obtains), a fortiori, one to which the ascriptions allegedly refer or in which the properties ascribed are realized. To be sure, there are "mental acts" (to use Geach's term) in which there is some particular internal occurrence (when a person is struck by a thought, talking to himself, or having mental images), but it is clear that this sort of phenomenon cannot carry the explanatory burden that is intended for mental representations in general. In particular, it could explain neither action nor a subsequent grasp of a language. This brief sketch needs filling out, but it provides an indication, at least, of how someone might begin to resist the idea that ascriptions of propositional attitudes function to pick out internal states of the individual, and that when "mental acts" do occur, they can provide the explanatory power required of mental representations.

3. THE THIRD chapter introduces the reader to simple computational devices to illustrate an example of a causal mechanism that "processes" or "computes" representations in a systematic way. The difference between *instantiating* and *computing* a function is discussed, and it is argued that the latter is needed if we are to understand the idea that minds are partly computers; only if the mind is a system that performs computations on mental representations will we be able to distinguish the idea that minds are computers from the weaker view that a theory of mind may be modeled on a computer. This distinction is crucial, since only the stronger would require us to posit causally efficacious, content-bearing mental states over which computations are performed. Crane goes on to consider whether machines can think and considers Dreyfus's objection that thinking requires *know-how* and cannot simply consist in rule-governed symbol manipulation; know-how requires the ability to form judgments about relevance. If such judgments were construed as higher-order rules, then it is unclear how it could be decided which of them were relevant to the case at hand without positing even higher-order rules, ad infinitum. Searle's Chinese Room argument is sympathetically considered and Crane admits that it is unlikely that something can think merely by manipulating symbols: a thinker has to be responsive to the meaning of the symbols and not just their shape. According to Crane, however, this does not tell against the idea that minds think partly by manipulating symbols, or by processing mental representations, and he points to familiar patterns of systematicity, consistency, and rationality in our thought processes as prima facie candidates for rule-governed processes. This leads to the suggestion that some representational states of mind (in particular propositional attitudes) are related to one another in the way that representational states of a computer are related: they are processed by means of algorithmic, or sometimes heuristic, rules. Set against a careful and sympathetic discussion of Fodor's Language of Thought (LOT) hypothesis, the computer analogy is invoked to do double work. First, it allows us to explain how mental representations (sentences of Mentalese) might be physically realized in the neuronal structure of the brain (as an English sentence might be stored in a computer). Second, it allows us to see how a brain might perform computational operations by mirroring the syntactic properties of sentences/thoughts so that semantically interpretable properties (like truth-preservation) will obtain in virtue of these causal structures (just as the computer processes symbols in accordance with their syntactic properties).

But the admission that manipulating symbols is not sufficient for thinking is more damaging than might be supposed. The argument that minds process mental representations is an argument that is based on the analogy with computers. We have an idea of what a mind's processing mental representations would involve, so the argument goes, because we know what it is for a computer to process representations. But it turns out that processing representations (construed as "mere" symbol manipulation) is not sufficient for thinking. To reject the analogy on the grounds that computers do not contain real representations would be, to quote Dennett, "[to discard] one of the most promising conceptual advances ever to fall into philosophers' hands."[2] To take the analogy seriously, however, would seem to lead to an embarrassing result: if the mind processes representations in the same way that a computer does, and if a computer's processes are not sufficient for thinking, then neither are the mind's. But it was the systematicity, consistency, and rational patterns displayed in *thinking* that led us to suppose that the operations over mental representations are rule-governed.

One response to this embarrassment (the one Crane opts for) is that *something else* is required for thinking, which partly consists in symbol manipulation. Presumably, the "something else" required is that the thinker be responsive to the meaning of the symbols and not just their shape. But now there is a danger that "being responsive to meaning" requires interpretation, and this is not allowed since mental representations were introduced (in part) to explain interpretation. Crane himself concedes that the supposed "interpretability" of the formal properties of *mental* representations gives rise to the "homunculus" fallacy: although we can make sense of the interpretability of a computer's representations by supposing that a programmer, say, is there to do the interpreting, we cannot suppose this to be the case for Mentalese. So, if we were to stick to the analogy with computers, we would have to posit homunculi to do the interpreting, which would lead to a vicious regress.

Crane agrees with Searle in rejecting Dennett's proposal for avoiding the regress problem (the proposal is to suppose that the homunculi posited become stupider and stupider until they finally drop out of the picture) on the grounds that even at the basic "switch" level, the position of the switches is semantically interpretable and would still depend on a programmer or designer to do the interpreting. Instead, setting up the discussion for the final chapter of the book, Crane suggests that the way to solve the problem is to say that mental representations get their meaning in a different way from other representations (i.e., they are not interpreted and thus not subject to the homunculus fallacy). The problem with this move, however (which Crane mentions but does not

---

2. Daniel Dennett, *Brainstorms—Philosophical Essays on Mind and Psychology* (Montgomery, VT: Bradford Books, 1978), 102.

develop), is that as long as one is willing to admit that there are certain alleged representations (viz., the sentences of Mentalese) that get their meaning in a way that does not involve interpretation, why not just say the same about public-language sentences or signs? If we were able to say the same about public representations, this would at least liberate the investigation of thought and propositional attitude ascription from the requirement that the representational power of the mind must be explanatorily basic.

4. Before leaving the Language of Thought hypothesis, Crane considers Searle's objection from "rule-following," which is credited to Quine's criticism of Chomsky. The argument is that the LOT hypothesis is committed to saying that our tacit knowledge of the rules of Mentalese and our (tacit) following of these rules is supposed to explain the systematicity and overall rationality of our thought and reasoning processes.

Quine's criticism is that with the introduction of *tacit* knowledge, it is difficult to see how one can make sense of the difference between genuinely following rules and merely acting in accordance with them. If merely acting in accordance with them were sufficient for thinking (or for language use), then there would be no reason to suppose that the rules are represented within the system at all, and the Mentalese (or innate grammar) hypothesis would collapse. Saying that the rules governing the operations of Mentalese *are* represented raises the specter of regress in two different ways. First, it suffers from the "homunculus" problem discussed above, which is that the very concept of representation seems to require that there be someone external to the system of representations (an "exempt agent") who uses or interprets it. This is a feature of representation in general and is simply inherited by the demand that the rules that govern the operations over (first-order) representations are also represented. The second regress arises because the contents of the representations now under consideration are supposed to consist in the alleged rules governing computations over lower-level representations. This is a problem, because if we require the existence of rules to govern the computational relations between representations, and if we suppose that these rules are themselves represented, then we shall presumably need higher-order rules to tell us how these rules-qua-representations are to operate, and so on. (This is a more general version of the "relevance" regress acknowledged earlier.) We could avoid the regress by denying that the rules need to be represented, but we would then run the risk of collapsing the distinction between computing and instantiating a function that is needed to sustain the LOT hypothesis. If we are to distinguish the claim that the mind is (partly) a computer from the claim that a theory of mind might be modeled on one (or if we are to distinguish the LOT hypothesis that the mind performs computations over representations from the significantly weaker claim that the thoughts we attribute to an individual accord with rationality constraints or principles) then it looks as if the rules governing the operations need to be represented.

The common move to make here (and, indeed, the move Crane makes) is to assume that the regress only arises if the rules are represented explicitly, and not if they are represented implicitly. But even if this could solve the first regress problem, it could not possibly solve the second. If rules governing the relations between representations are needed in the first place (to explain the systematicity or rationality of thought processes), then rules governing relations between the rule-representations will be needed as well. This second regress arises because rule-representations are invoked *period:* it is oblivious to whether they are represented explicitly, implicitly, or tacitly.

Moreover, it is not even clear how turning to implicit representation could solve the first regress problem, that representations of any kind require an interpreter. Presumably, explicit representation is thought to be the culprit because it requires actual interpretation, whereas implicit representation requires merely that the causal processes within the system are *interpretable* as rule-governed. But Dennett's proposal for dealing with the homunculus problem was rejected on the grounds that even basic symbols require interpretation; the proposal for the construal of the representation of rules as *implicit* should be rejected for the same reason.

If what is interpretable is to be interpreted, an interpreter is still required to do the interpreting. If, however, the rules are only representational in the weak sense that the causal processes may but need not be so interpreted, then this is to admit that the rules' representation is not necessary and that the uninterpreted causal processes are sufficient for explanation. This is tantamount to (re)-collapsing the rule-following/rule-conformity distinction and, again, is something that the computationalist cannot accept.

Perhaps it is time to consider why a theorist needs to introduce rules into an account of language, action, or thought in the first place. Something like principles or standards seem to be needed in order to capture the normative nature of language, rationality, and thought. Appealing to the meaning of a word or to norms of rationality allows us, for example, to pinpoint what has gone wrong when a language user misunderstands a word or when a rational agent acts irrationally. Now, if an individual's knowledge of meanings or principles of rationality were thought to be a matter of his having internalized rules that are action-determining, then we would be unable to diagnose this special kind of error ("internal" error, which allows that the one who commits it is still participating in the practice that the rules govern). Somebody who makes mistakes in inference is still a thinker; someone who mistakes the meaning of a word might still say something meaningful; someone who acts irrationally is still acting in a way that is rationalizable. The phenomena to be explained (thinking, action, language) allow "internal" mistakes of this sort: mistakes that do not completely strip the phenomena of their credentials. If the norms that must somehow feed into an account of the systematicity or rational patterns of thinking were simply to be read off a person's thought processes (assuming this makes sense), then we would not have the necessary logical space that is needed to make sense of irrational or nonlogical steps in thinking. To put it in a slightly different way, if the individual is capable of deviant thought patterns (for example, mistakes in inference), then her thought patterns had better not determine the alleged rule in accordance with which the patterns are governed, or the rule will sanction these deviant patterns. If the individual is not capable of mistakes in inference, or other "deviation" in the systematic processing of thoughts, then we are working with a different notion of "thinking" from that which the computational view was introduced to explain. In any case, in the absence of error of this kind, there would be no need to suppose that there are norms or standards that are somehow represented in the mind of the individual whose thought patterns are measured against them. Just as we suppose that there are laws of nature to which falling bodies conform without supposing that these are represented as rules and followed by the falling bodies, so might we suppose that there are patterns or regularities in thinking to which thinkers conform without supposing that these are represented as rules which they follow.

Standards or norms allow us to calibrate the relative "success" of the phenomenon under consideration. It is tempting to try to explain an individual's ability to act in a way that can be calibrated by these norms by supposing that they figure as the content of rules he follows. Yet if knowledge of the rules determines or necessitates that the individual will act as they prescribe, then we are left without the possibility of "internal" error. The only plausible way forward is to relax the requirement that knowledge of norms is necessary for crediting an individual with the ability in question (sometimes mere conformity will suffice) and, when it is required, to relax the requirement that knowledge of the norms necessitates that the action will conform. But these moves are of no help to the mental representationalist. If mere norm-conformity is to be an acceptable account of what it means to say that mental representations are governed by computational rules, then he might as well have allowed mere conformity when it came to explaining an individual's behavior (or in saying what the meaning of a sign consists in) in the first place. Nor, if knowledge of the norms is to play some sort of role here, can it be such as to necessitate or determine that action or thought will conform; if it necessitates anything, it is that action or thought *should* conform. But it is difficult to make sense of this prescriptive role when knowledge of the norms is thought to be embodied in computational processes realized in the brain. Although Crane does not make explicit the link between this normativity objection and the distinction between rule-following and mere rule-conformity, he does concede that the rules governing Mentalese processes would have to be construed as nonnormative. Citing Fodor, he offers two responses on behalf of the representationalist. The first, "defensive," response is that the Mentalese hypothesis is not committed to saying that the (admittedly normative) rules of logic or the norms of practical reasoning are the rules that govern it. The rules that govern Mentalese are to be discovered by empirical scientific investigation. The second, "aggressive," response is that even if the norms of practical reason or the rules of logic were discovered to be the rules governing Mentalese, then

> they would be idealizations from the complex, messy, actual behavior of minds. To state the rules properly, we would have to add [a ceteris paribus clause]. But this does not undermine the scientific status of Mentalese, since ceteris paribus clauses are used in other scientific theories too.[3]

The first response is a dodge. The computational theory of mind was introduced to explain the kind of systematicity or rationality exhibited in certain thought patterns that the appeal to principles of commonsense psychology and logic made vivid. It was this type of systematicity that was supposed to be in need of explanation; presumably it still is, even if Mentalese is suddenly whisked away to explain something else, such as the sort of systematicity yet to be discovered by empirical investigation. If the rules (whatever they express) are merely read off a person's behavior, then they will not allow the logical latitude for mistakes (or for moves that fall short of the standard). If there is no possibility of mistake, then there does not seem to be any need to appeal to a normative notion such as *rules.* And if there is no need to appeal to such a normative notion, then there is no need to effect an internal, cognitive relation between these standards and the behavior they are thought to govern.

The suggestion that the rules of Mentalese are *idealizations* from the behavior of actual minds treads suspiciously closely to the idea that the norms (of language, or rationality, etc.) govern *practices* in which individuals participate. The comment about ceteris paribus clauses takes us quickly back, however, to the realm of science. But with science as our paradigm, we are left with patterns of behavior that are explained

3. Crane, *The Mechanical Mind,* 153, and Jerry Fodor, *Psychosemantics—The Problem of Meaning in Philosophy of Mind* (Cambridge, MA: MIT Press, 1987), chapter 1.

or not insofar as they succeed or fail to conform to general regularities. The notion of "internal" error, however, is not captured by this simple distinction between conformity and nonconformity and the need for rules is, again, unclear.

5. THE FINAL, most difficult, chapter brings us up to date with contemporary attempts to naturalize content. When it comes to giving a naturalist account of content itself and not just a causal (computational) account of the operations that govern states that exhibit content, the problem of error is taken much more seriously. Crane argues that the naturalist's attempt to state causally necessary and sufficient conditions for a thought to represent (or reliably to indicate) an object, for example, is completely disabled by its failure to accommodate the fact that our thoughts (percepts, etc.) can *misrepresent.* Reliable indication cannot be a necessary condition of representation, or misrepresentation would be ruled out; reliable indication cannot be a sufficient condition of representation, or systematic error would force us to allow unacceptable disjunctive representations. In either case, the possibility of error is not properly accommodated. Indeed, Crane himself suggests rejecting reductionist accounts of content because they do not accommodate the problem of error. In a nice, if somewhat ironical, summing up of the problem, he writes:

> Philosophical attempts to explain the notion of representation by reducing it have not been conspicuously successful. They all have trouble with the problems of error. This is unsurprising: the idea of error and the idea of representation go hand-in-hand. To represent the world as being a certain way is implicitly to allow a gap between how the representation says the world is and how the world actually is. But this is just to allow the possibility of error. So any reduction which captures the essence of representation must capture whatever it is that allows for this possibility. This is why the possibility of error can never be a side-issue for a reductive theory of representation.[4]

4. Crane, *The Mechanical Mind,* 196–197.

Now it is an interesting fact that the naturalists who attempt to give constitutive accounts of perceptual representation or thought content have taken the problem of error very seriously.[5] It is also interesting that the problem of error is somewhat less appreciated (for some reason) when the theorists are engaged in naturalizing meaning or the ostensible rules that govern thought.[6] The problem of error that arises for accounts of representation is the same "normativity" problem that arises for meaning-rules and for the rule-governed computations of Mentalese, except in the latter cases the required logical slack will have to be found between the rules and the operations they govern, instead of between the representational contents and the objects or states of affairs they represent. The move to dismiss the possibility of error regarding the operations over representations is not a move available to theorists about meaning or content: the possibility of error (in the guise of a possibly deviant interpretation) is what motivated us to introduce interpretation into the analysis of (public) representation in the first place. If we can dismiss the fact that sometimes a person (who can still be credited with thinking) does not always think in accordance with the norms of rationality or of logic as irrelevant to our account of the operations that govern thinking, then we could have dismissed the fact that a deviant interpretation of a sign is available in our account of what it is for that sign to represent an object.

The "non-reductive" view that Crane tentatively embraces goes something like this.[7] Those impressed by the ostensible explanatory power of the computational theory of mind might plausibly resist the challenge to give an independent account of the representations that

5. See Fred Dretske, "Misrepresentation," in *Belief: Form, Content, and Function,* Radu Bogdan, ed. (Oxford: Oxford University Press, 1986); Jerry Fodor, *A Theory of Content* (Cambridge, MA: MIT Press, 1990) and *Psychosemantics.*

6. See Fodor, *Psychosemantics;* Paul Horwich, "Meaning, Use, and Truth," *Mind* 104, 414 (1995): 355–368. Consider also Chomsky's remark that there are no arguments that he knows of for irrationality in *The Chomsky Reader,* James Peck, ed. (New York: Pantheon, 1987).

7. Crane credits this primarily to Robert Cummins, *Meaning and Mental Representation* (Cambridge, MA: MIT Press, 1989), and Frances Egan, "Individualism, Computation, and Perceptual Content," *Mind* 10, 403 (1992): 443–459.

figure in computational relations. The inner states of an organism that enter into formal or syntactic relations must indeed have a specification that is independent of their representational properties, but the representational properties themselves do not have to be construed independently of the computations attributed to the organism. What it is to treat these internal states as representations is simply to specify a mapping or an "interpretation function" from the state formally described to an abstract representational content. According to Crane, this allows us to agree with Fodor that there is "no computation without representation."

> But this does not mean that we need to give a reductive account of what a representation is. Representation is just another concept in the theory; it does not need external philosophical defence and reduction.[8]

To make the interpretation-computational view vivid, an analogy with numbers is used.[9] Just as we use the number 2.2 to pick out the weight of a bag of sugar (say, in pounds), whose arithmetical operations (involving other numbers) "mirror" specific physical relations between specific weights, so, too, do we use sentences to pick out a person's mental states in the course of "measuring" her behavior. Once we have identified a mental state by mapping it onto a sentence, we can see that the logical relations between the sentences "mirror" the psychological relations between the mental states.

Crane notes two consequences of taking this analogy seriously. The first is that, strictly speaking, numbers are not necessary to characterize the weights of objects: we could, for example, track the weight of an object simply by comparing it to the weight of another. This feature of the analogy, Crane thinks, should not be a problem for the interpretation-computationalist, since he, too, is committed to the idea that

---

8. Crane, *The Mechanical Mind*, 192.

9. Crane credits the analogy to Field, who in turn credits it to David Lewis. Note that it is an analogy that has also been employed by Davidson and Churchland.

the states that participate in computations have a formal description not given by the sentences that "map" their content. The second feature, which Crane does find troubling, is that nothing ensures that the interpretation function will express a one-to-one mapping, as opposed to a one-to-many mapping. This is certainly the case when numbers are used to measure objects: the choice of numbers depends on the unit of measurement. But if there are many interpretation functions that will assign distinct interpretations to the symbols, and the choice between them is determined only according to which has more explanatory power, then "it looks like this explanatory power is an illusion, since it is purchased at too cheap a price." The choice that we are left with is either to accept the "implausible consequence that mental states do not have unique contents which are essential to them" or to find other, naturalistic, constraints that determine which interpretation function is the right one.[10] If such a constraint can be found, however, the idea that the representational content is the result of an interpretation function is threatened with superfluity. Crane leaves the debate here, but suggests that a way out would be to "bite the bullet" and deny, what seems to him to be a common-sense view, that our beliefs have unique contents that determine them. Strictly speaking, the computational view is an empirical hypothesis about nonconscious states, and cannot be overturned on the grounds that it is counterintuitive.

Thus we end up with a strange—and unstable—hybrid view that consists in a strict realism about the vehicles of mental representation, which have full causal power, together with the eventual acceptance of an antirealism about content itself, insofar as content turns out to be explanatorily unnecessary, indeterminate, and interpretation-dependent. But this cannot be acceptable. The notion of interpretation cannot be redeployed here without running directly foul of the circularity that the naturalist has, all along, been taking such pains to avoid. Even if interpretation is something, to paraphrase, that a "computational theorist's mind bestows on internal states," then mental representations turn out not to be basic after all. They can hardly be introduced to *explain* the interpretation

10. Crane, *The Mechanical Mind,* 195.

of other symbols. The defenders of the view seem to think that it is saved from total collapse by the insistence that the internal states that are the bearers of content are independently specifiable in virtue of their formal or syntactic properties, even though their representational properties depend upon interpretation. But it is simply not an option for the representationalist to demur from the naturalistic program at the last minute and admit that one can, as it were, calibrate the behavior of the individual to be explained in the *absence* of the representational properties of the alleged internal states; it was because of their alleged representational properties that we were led to posit the internal states in the first place. If content is, strictly speaking, unnecessary, then surely the *bearers* of content are equally so. Thus, the battle cry "No computation without representation" has to be reserved for the thoroughgoing realist and cannot be adopted by the antirealist about content. It has to mean, that is, that representation is *necessary* for computation, not that we can interpret the formally described internal states of the organism as representations if we want to. For anyone willing to go "antirealist" now might as well have gone "antirealist" a long time ago, and refused to accept that the ascriptions that calibrate the behavior of the individual must express properties that are realized in any internal state of the individual, let alone in a causally efficacious, physical one.

Perhaps the "interpretation-function" that is introduced at the end is not supposed to depend on the availability of an actual interpreter. Perhaps the important point here is the notion of a *function* or *mapping* which only fails to adhere to the strict requirements of realism by failing to determine a unique content. The problem, though, is that it is unclear how this proposal is supposed to avoid the problem of error that plagued the reductionist attempts. The problem of error reoccurs on this view if we ask for details of the "interpretation-function" that maps internal states onto representations. If the function gives us a necessary condition for mapping an internal state onto a representation (or, allowing for the possibility of having different "units of measurement," onto a certain disjunction of representations), then it would seem that misrepresentation is impossible here, too. If it gives us a sufficient condition for mapping an internal state onto a disjunction of representations, then this disjunction of representations will not only be constrained by different "units of measurement," but will allow in all sorts of deviant, but possible, interpretations too. It is true that the problem of error will thwart any attempt to read the rules off an individual's (or for that matter, a group of individuals') behavior, on the grounds that deviant behavior will turn out to be sanctioned by the rules. But even if the account is nonreductionist, I suspect there will be a problem with any attempt to fashion the norms governing a practice into a constitutive account: an account, that is, in terms of necessary and sufficient conditions of what it is for a move to be a move within that practice. When we try to describe the norms that govern a practice, we are abstracting successful from unsuccessful moves; rational from irrational actions; misperceptions from veridical ones; and successful from unsuccessful steps in thinking. The distinction is between two different kinds of moves within a practice—between different kinds of actions, or perceptions, or steps in thinking—not between what is a move in the practice and what is not.

Most of those investigating language and action will claim that they are not interested in giving a "constitutive" account. They are interested in an empirical investigation of the psychological laws that govern the phenomena. These laws will no doubt be hedged with ceteris paribus clauses and will admit of exception just like most laws of the special sciences. Or, even if the investigation is focused on the conceptual landscape delineated by the appropriate application of mental terms, the sought-after generalizations will at best give defeasible conditions for the application of the relevant concepts. I have already noted, regarding the possibility of psychological laws, that this is not the sort of error that an appeal to rules makes possible. Second, any proposal for relaxing the standards of explanation to allow exception clauses (or defeasibility conditions) which themselves may never be noncircularly spelled out leaves itself open to the criticism that if such a move is available now, it was available right at the beginning when we were attempting to give an account of public representations. It was the search for a constitutive account of public representations that motivated the introduction of mental representations in the first place.

To be sure, there have been other motivations in the history of this debate for positing internal representational states. The fear of dualism has led many to insist that content, or at least its vehicles, must have some sort of physical realization, and a corresponding search for scientifically respectable (i.e., causal and possibly nomological) explanations has cast its own constraints on the kinds of events that are eligible to enter into causal relations. It is, indeed, a particular idiosyncrasy of Crane's book that he refuses to be drawn into the traditional debate between physicalism and dualism and will not use physicalism to motivate the introduction of mental representations.[11] Does this strategy work? The problem, I suspect, is that he does not go far enough in rejecting the distinction. Too much of contemporary philosophy of mind, in general, including the particular aspects that Crane accepts here, is framed on the premise that dualism is a serious threat. What else, for example, if not an implicit commitment to physicalism, would lead someone to be sympathetic with the view that thoughts, qua sentences of Mentalese, are stored in the neuronal structure of the brain or to be satisfied if it is the formal or syntactic properties of the alleged bearers of content that end up doing the causal work? In refusing to be drawn into this debate, Crane seems somewhat unfairly to avoid the important charges of "epiphenomenalism" that are being leveled currently against those mental realists who think that they can *also* meet the demands of physicalism. In the discussion above, I have tried to show that explanatory redundancy of a kind is still a problem even if physicalist motivations are pushed into the background.

What, then, of the intuition that beliefs and desires (whatever these turn out to be) are causally efficacious? This topic needs more discussion than I can give it here, but it should be clear from the discussion so far that if we accept the idea that causes necessitate (or even probabilize) their effects, the idea that reasons are causes is not likely to dovetail nicely with the idea that reasons exert a normative influence over the one who accepts them.[12] Nobody, of course, wants to deny that the

11. As does, for example, Hartry Field, in "Mental Representation," (1978), in *Readings in Philosophy of Psychology*, vol. 2, Ned Block, ed. (Cambridge, MA: Harvard University Press, 1981).

12. See the chapters in part 2 of this volume as well as "Ryle's Regress and the Philosophy of Cognitive Science" (chapter 12) for more discussion.

point and rationale of ascriptions of beliefs and desires is to enable us to understand the individual to whom they are attributed; nor should it be denied that such ascriptions are necessary for explaining human action. One might agree that they are more sophisticated tools of measurement than, say, their cruder character-trait cousins. But just as ascriptions of character traits might afford some degree of explanation by identifying a particular action as part of a more general pattern without supposing that the character traits themselves are realized in some internal state of the individual to whom they are ascribed, so too with the ascriptions of propositional attitudes or reasons, in general. If something along these lines is correct, there does not seem to be much reason left to believe in the existence of mental representations.

CHAPTER TEN

On the Conceptual, Psychological, and Moral Status of Zombies, Swamp-Beings, and Other "Behaviorally Indistinguishable" Creatures

1. "Philosophical zombies," writes Robert Kirk, "are exactly like us in all physical respects, right down to the tiniest details, but they have no conscious experiences. My zombie twin not only looks, behaves, and is disposed to behave just like me, he is a perfect particle-for-particle replica. Naturally he gets treated as if he were conscious. . . . However, this is a philosophical example, and this particular physical replica is *defined* as not having any conscious experiences: "all is silent and dark within."[1] Zombies of philosophical lore are thus very different from the "living dead" zombies of horror films and even from humanlike creatures of science fiction whose behavior is oddly disturbing in ways that might give us pause. Philosophical zombies, as Kirk has defined them, are indistinguishable from humans in the sense that there is no way, under any possible tests, of telling them apart. The only difference is something that is supposedly impossible to discover: they lack conscious experience.

1. Robert Kirk, "Why There Couldn't Be Zombies," *Proceedings of the Aristotelian Society,* supp. vol. 63 (1999): 1–16.

Zombies are presently generating much discussion in the philosophy of mind and consciousness studies.[2] For if a creature could be physically, functionally, and behaviorally indistinguishable from humans (in the rich sense implied) yet lack conscious experience, then the theories of mind that tie the nature of the mental too closely to physical, functional, or behavioral conditions will seem to have left something crucially *mental* out of their theories. If having conscious experiences is necessary for being conscious—as these discussions assume—then the theories that cannot accommodate it will fail as theories of mind.

Kirk defines zombies as our physical duplicates, but recent literature has wanted to distinguish this variety from our functional and behavioral zombie twins. The increasing complexity of these discussions results not only from this tripartite distinction, but also from its modal permutations: thus more focused disputes continue about whether our behavioral, physical, and functional zombie twins are or are not conceptually, naturally, or metaphysically possible. With the conceptual possibility of behaviorally indistinguishable zombies considered the weakest—and generally uncontroversial—case, the discussions tend to concentrate on the natural or metaphysical possibility of physically or functionally indistinguishable consciousness-deprived creatures.[3]

Those attempting to deny the weakest claim that behaviorally indistinguishable consciousness-lacking creatures are conceptually possible

---

2. See, for example, Robert Kirk, "Sentience and Behavior," *Mind* 83 (1974): 43–60; "Zombies v. Materialists," *Proceedings of the Aristotelian Society,* supp. vol. 48 (1974): 135–152; Daniel Dennett, *Consciousness Explained* (London: Penguin Books, 1991); Todd Moody, "Conversations with Zombies," *Journal of Consciousness Studies* 1, 2 (1994): 196–200; David Chalmers, *The Conscious Mind: In Search of a Fundamental Theory* (New York and Oxford: Oxford University Press, 1996); and Daniel Dennett, "The Unimagined Preposterousness of Zombies," *Brainchildren—Essays on Designing Minds* (London: Penguin Books, 1998), 171–177.

3. Indeed, the main reason, it would seem, for distinguishing the modal claims is to fine-tune a disagreement between functionalists and type physicalists about whether functionally identical zombies are possible. If they are, this would seem to give some support to the type physicalists. See Tom Polger, "Zombies," in *A Field Guide to the Philosophy of Mind,* Marco Nani and Massimo Marraffa, eds., Societa Italiana Filosofia Analitica (2001) http://host.uniroma3.it/progetti/kant/field/zombies.htm (accessed 1 November 2001).

are promptly accused of an implicit commitment to "logical" or "analytic" behaviorism. Since this doctrine about the meaning or translatability of statements containing mental predicates can be dismissed out of hand, the conceptual possibility of behaviorally indistinguishable zombies is left unchallenged by most players in the dispute. Dennett is an important exception. In calling the very idea of zombies "utterly preposterous" he remarks that we—all of us—are more committed to behaviorism than we would like to believe. Dennett is on to something, but it is important not to overstate the case. In particular, it is important to see that a rejection of the possibility of behaviorally indistinguishable consciousness-lacking creatures does not carry the burden of a hopeless semantic theory. This, in any case, is what I hope to show.

In this paper, I shall argue that we should reject the idea that a creature that is in the same circumstances as his (conscious) human counterpart and behaviorally indistinguishable from him could yet not be conscious. I shall also urge rejection of the broader claim that such a behavioral twin could lack a mind. I understand "behavioral indistinguishability" to entail that there is nothing from the skin outwards, as it were, by which we could distinguish these creatures from their human counterparts. They share facial expressions, involuntary reflexes, as well as behavior construable in the richer sense as speech and action, and they would manifest behavior that is indistinguishable from their human counterparts, considered in the same circumstances, under all possible tests. Unusually for discussions of this kind, the argument will not appeal to philosophers' unreflective responses, intuitions, or imaginings. Nor will it depend on a prior commitment to metaphysical claims or to what is or is not consistent with the laws of nature. Most importantly, the arguments of this paper do not rely on an adherence to logical behaviorism in any of its familiar forms. I shall argue on psychological/explanatory grounds that the application of consciousness concepts cannot legitimately be withheld from creatures that are behaviorally indistinguishable from us. I shall argue on moral grounds that propositional attitude concepts cannot be withheld from such creatures either.

A simple argument to the effect that zombies are conceptually impossible would tend to have little impact since the people proposing theories in philosophy of mind today do not take themselves to be offering analyses of our mental concepts whose integrities need to be (wholly) preserved. This is because the attempt to construe mental phenomena as part and parcel of the natural world is often thought to necessitate reductions of some kind.[4] When the proposed reductions fail to capture certain features of our common-sense mental (or semantic, or other normative) concepts, we are invited to accept a change or variation in the application of these concepts in the light of the proposed revision, which is usually defended on "explanatory grounds." My strategy will therefore be not only to show that our common-sense consciousness and propositional attitude concepts *are* properly ascribable to creatures that are behaviorally indistinguishable from us, but also to urge that the invitation to change these concepts in the ways envisaged would not only distort their explanatory power, but so, too, their ethical import.

My argument begins by noting that creatures that are behaviorally indistinguishable from their human analogues will meet all the second- and third-person criteria for the ascription of "experiential" mental predicates in exactly the same situations as their counterparts do. These predicates will include the sensation concepts of seeing, hearing, touching, smelling, and tasting; they will include those containing the term "conscious," itself; and they will include the related concepts of the feelings and emotions. In fact, one needs something *far less* than behavioral indistinguishability in order to ascribe these concepts to a creature: after all, we ascribe many of them to animals. Our behavioral twins will also, however, possibly unlike animals, warrant the ascription of concepts associated with the more complex notion of "self-consciousness."

---

4. These might involve, for example, the search for theoretical identities which would form part of the attempt to provide bridge principles (definitions or empirically certified correlation laws) via which the ostensible theorems and axioms of a commonsense psychological "theory" can be derived from a base theory whose theoretical vocabulary does not include mental or semantic terms. Or such a "naturalist" might eschew the search for contingent identities demanded by Nagelian reduction (and the Hempelian, deductive, nomological model of explanation upon which it rests) in favor of a functional reduction. See Jaegwon Kim's discussion of this choice in *Mind in a Physical World* (Cambridge, MA: A Bradford Book, MIT Press, 1998), chapter 4.

As a way of assembling reminders of how our consciousness concepts are applied and defended, let me introduce the reader to zombie John. If zombie John is behaviorally indistinguishable from human John then, when sitting on the rooftop terrace in Provence, sipping a glass of wine, he will be just as competent as human John to describe, for example, the way the sunlight colors the rocky face of the Massif de la Sainte-Baume, or the buttery flavors of the white wine from southern Burgundy, or the difference in the resonance of the bells as they strike in the two village clock towers. In normal circumstances, we ought to be able to cite these abilities to defend our application to him of the concepts of seeing, tasting, and hearing.

Not only will zombie John satisfy the criteria for the second- and third-person application of these concepts, he will have no more and no less difficulty than human John in applying these concepts to himself. His application of these concepts to himself should be defeated or allowed to stand on the same grounds as human John's. If, for example, I doubt whether zombie John is really able to see La Sainte-Baume, then I might ask him to point to the meteorological station nestled on one of its peaks. If zombie John is behaviorally indistinguishable from human John then he will be able to do this as soon as he puts on his glasses. But if we are told that zombies are *defined* as both being behaviorally indistinguishable from us and as not being able, for example, to see, then we should insist that we no longer understand what it means to see, and so forth, for all the other terms associated with consciousness concepts that have a perfectly normal role in our language.

Suppose for the sake of argument that zombie John lacks something that human John has: something only accessible privately. Notice how difficult it would be to say what it is that he lacks. For every time we attempt to name something: a feeling, a sensation, an experience, or a perception, zombie John will satisfy us exactly as well as human John by his way of talking, his discriminatory abilities, and by other sorts of responses, that we are correctly attributing these concepts to him and that he is correctly attributing them to himself.

In any case, even if we could make sense of the idea that there is something (ineffable) missing in the case of zombies, it is not obvious how this ought to affect our use of the relevant expressions. This can be seen by confining our attention, for the moment, to humans. After all, if *human* John were suddenly to declare that "all is dark within" but the rest of his behavior remained the same, we would be very puzzled. After giving him various tests to determine, say, if he is blind, deaf, and cannot smell or taste, we would be obliged to point out that he sees, hears, smells, and tastes as well as before. If he goes on to *insist* that it is still "dark inside" even though he continues to stamp on the terrace every time a pigeon lands on the roof, whistles along with the radio to the principal theme of Shostakovich's first cello concerto, and can even distinguish the chardonnay of Puligny-Montrachet from that of Chassagne- Montrachet, we may very well be at a loss for words. But we would not, I maintain, conclude that he is not conscious or that he is not having sense experiences. This is important, because it illustrates, as does the case of blindsight, that when first- and third-person criteria for the application of consciousness concepts pull apart, this may result in some hesitation about how to apply them. But first-person denials would not result in the simple rejection of their application and, in any case, for the zombie there will not be any first-person denials.[5]

Nonetheless, there may linger an irresistible temptation to suppose that there might still be something that a zombie lacks but we have even if it is something that cannot be identified, named, or described—at least not in our "public" or "common" language. We can imagine what it is like for us in the dark, for example, and it seems we can separate this in our minds from the discriminatory abilities that we have when it is light and we can see. Thus it seems we can imagine that someone could exercise these abilities and yet still experience the world as dark. I do not have to argue against this. It is enough for my purposes to point out that this imaginable "darkness" could not affect the meaning of terms like "conscious," "experience," "sensation," and

5. The term "blindsight" illustrates the point nicely: the subject's reports *not* to see are accommodated by calling the phenomenon "blind" and the subject's ability to pick out the object nonetheless is accommodated by counting it as a case of "sight."

so forth, simply because the light with which it is supposed to contrast does not, because it cannot (ex hypothesi), enter into the justifications or explanations we give for the application of those words.[6]

Philosophers could, of course, invent a term—like "qualia"—for what it is that we have but zombies lack, but then, of course, nothing follows from this about whether they have sensations or are conscious in the normal senses of these terms. If the application of consciousness concepts to zombies is refused, the hidden assumption must be that whatever is alleged to be missing from their experience but present in ours determines the meaning of our consciousness concepts—else no one would be able to reject the application of such concepts to zombies because of what they are supposed to lack. But if whatever is allegedly missing in the case of zombies but present in the normal case does not (because it cannot) figure in the explanations and justifications we give for the application of consciousness concepts, then it cannot determine their meaning. If it cannot determine their meaning, then by what right do we deny their application to zombies? For we could defend our right to apply the concepts to them in the same circumstances as we would defend it for us.

If zombies are "behaviorally indistinguishable" from us then presumably they fall asleep, wake up again, and are susceptible to blacking out when traumatized. As we use these terms normally, we would, with perfect right, be able to say about zombie John that "he is not conscious of the mess being made in the kitchen because he is asleep upstairs" or "he has not been conscious for two days since the binge." If zombie John is never conscious, then how would we describe these

6. I have claimed that whatever is allegedly present in our case but missing in zombies does not (because it cannot) enter into the justifications and explanations given for the application of consciousness or sensation terms. So even if whatever is allegedly present in our case could be given a private definition, it could not infiltrate public meanings. A more ambitious claim would be that there could be no private definition of whatever is allegedly present in our case but missing for zombies on the grounds that no genuine justifications and explanations could be given for its application. A still more ambitious claim would cast doubt on the existence of a determinate private something, which is alleged to be present in our case but missing from zombies. Although I am sympathetic with them all, the arguments of this chapter rely only on the first, least ambitious claim.

circumstances? It cannot simply be that he *seems* to be awake and thus *seems* to be conscious when he gets over his hangover and emerges from bed, for if he is indistinguishable from human John, then we shall need to reserve these descriptions for when he is sleepwalking. (It cannot be that he is always sleepwalking, for we shall need to reserve this description for those rare nights when he goes to bed and suddenly finds himself in the kitchen a few hours later.) If zombie John is indistinguishable from human John we shall in any case need to preserve "seems-talk" to describe what happens when zombie John is feigning.

These arguments may sound familiar to those who recall the philosophical discussions of the problem of other minds. Indeed, the possibility of zombies reintroduces the problem of how we could ever know, or be justified in assuming, that other minds exist. Someone might thus be tempted to misconstrue the argument of this chapter as one that mixes up questions about what one could possibly know with questions about what is the case. But the argument I am offering here does not run from epistemology to metaphysics. It is logically prior: it concerns meaning.[7] To accept the conceptual possibility of zombies would be to accept the invitation to stretch our consciousness concepts so that the normal explanations and justifications we give for their application no longer apply. There may be good theoretical reasons to revise our common-sense concepts in certain domains.[8] But what reasons have we to accept this invitation here?

I have shown how accepting this invitation would radically alter our application of the predicate "is conscious" as well as predicates containing sensation terms. Reflect, too, on the normal use of the concepts of feeling and emotion. If zombie John is behaviorally indistinguishable

---

7. Someone might also be tempted to interpret mine as the claim that the possibility of zombies would render consciousness epiphenomenal. See, for example, Dennett, *Consciousness Explained,* 405, and John Perry, *The 1999 Jean Nicod Lectures* (Cambridge, MA: MIT Press, 2001). But I am not theorizing about the (lack of) causal power of conscious states. Again, my argument is about meaning.

8. But see note 6, above, for how stronger claims than I defend in this chapter might challenge the very coherence of the particular revision proposed here for our common-sense mental concepts.

from human John then he will sigh when his beloved kisses him on the back of his neck when he is working at his desk, growl when she pours more wine for herself than for him, and roll his eyes (heavenward) when the natives compliment her French. These facts, in normal circumstances, would justify us in saying that he likes it when she kisses him on the back of the neck, that he is annoyed when he does not get his fair share of wine, and that he is amused by French *politesse.*

Or consider the concepts associated with self-consciousness. When human John sees pigeons he will often be aware of his seeing pigeons. When human John wants more wine he may be aware of his wanting more wine. To paraphrase Bertrand Russell (and turn his argument on its head), there is nothing, ex hypothesi, that zombie John is immediately acquainted with when he sees pigeons or when he wants more wine, but he will nonetheless satisfy the criteria for "self-consciousness" as well as human John. He will, that is, defend his right to say he sees the pigeons or that he wants more wine with comments like, "I know when I see pigeons or I know when I want more wine!"[9]

He will also be able to say, with perfect right, that he had not been conscious of the hurt he would cause at that memorable dinner at the beginning of their relationship by ignoring the food and regaling the wine. He could defend this right, or at least try, with the claim that he did not know how difficult the meal had been to prepare, or that he had not realized how eager its cook had been to please, or that, coming from him, the fascination with the wine was not meant as a criticism of the food.

2. Someone might question my grounds for using predicates containing "knows," "realizes," or "means that" to describe zombie John, in justifying the claim that he was not conscious of causing hurt when he enthused about the wine at the expense of the food. But the same kind of argument can be given to defend the right to use these terms as well as related ones like "believes," "understands," "desires,"

9. Bertrand Russell, *The Problems of Philosophy* (London: Oxford University Press, 1967), 26.

and "intends." The discussion, however, has become hindered because zombies are *defined* as not being conscious, and some think it follows that they cannot have minds at all. Since an example is needed of a creature that is behaviorally indistinguishable from his human doppelgänger but whose mental status is not ruled out by definition, let us switch examples and discuss another creature that has generated a great deal of attention. "Swampman" is, ex hypothesi and as a matter of coincidence, behaviorally, functionally, and physically indistinguishable from his human opposite.[10] But, he has a deviant history: he came into being spontaneously in a lightning storm. Although not *defined* as lacking mentality, Swampman is nevertheless believed by certain philosophers not to warrant the attribution of propositional attitudes.

Having been introduced by Davidson in an argument for content-externalism, Swampman has been resurrected in discussions, among others, about teleological theories of representation whose notion of *proper function* makes essential reference to the evolutionary history of inner representational states.[11] Insofar as their proponents are giving us semantic theories about the content of the mental representations that figure in a theoretical explanation of the propositional attitudes that our common-sense concepts are alleged (however roughly) to pick out, they are committed to viewing the ascription of propositional attitude concepts to creatures like Swampman as at best useful for practical purposes if not out and out mistaken. This commitment is recognized and accepted by proponents of the view.[12]

---

10. Many philosophers think that a particle-for-particle identical being will ipso facto be functionally and (some think thus) behaviorally indistinguishable from his human counterpart. This in general is supposed to follow from the acceptance of certain supervenience claims. Nothing about what I have to say in what follows commits me to these claims. It is enough for my purposes to be focusing on the stipulation that Swampman is behaviorally indistinguishable from his human counterpart.

11. Donald Davidson, "Knowing One's Own Mind," (1987), *Subjective, Intersubjective, Objective* (Oxford: Clarendon Press, 2001), 15–38.

12. See, for example, Ruth Millikan, *Language, Thought, and Other Biological Categories* (Cambridge, MA: MIT Press, 1984), 93, and David Papineau, "Representation and Explanation," *Philosophy of Science* 51 (1984): 565. For a similar line see Fred Dretske, *Naturalizing the Mind* (Cambridge, MA: A Bradford Book, MIT Press, 1995), chapter

In agreeing to delimit their application in the light of these theories, however, we would be doing unacceptable damage to our ordinary mental concepts. It may be thought innocuous to deny these creatures' mentality, but with mentality goes agency and personhood, and with agency and personhood, the whole framework that would render these creatures appropriate subjects of respect, dignity, moral praise, and blame.[13] The teleosemanticists' defense of their position seems to rest primarily on the "vanishingly small" physical possibility of such creatures.[14]

This defense, however, completely misses the point that our mental concepts are applicable to naturally impossible, wholly fictional creatures, and that this application itself can be defended on *explanatory* grounds. When human John risks his life by leaning over the rooftop terrace to fasten chicken wire over the opening between the terrace ledge and the roof below, his action can be explained by his hatred of pigeons. His hatred of pigeons can be explained by his conviction—correct or incorrect—that pigeon droppings contain acid and will eventually destroy the terracotta roof tiles. When Swampman John risks his (equally vulnerable) life by leaning over the rooftop and behaving similarly it cannot plausibly be considered a theoretical advance to be told that although his behavior is *interpretable* as the attempt to prevent pigeons from congregating in the gap between roofs, it is

5. See also their contributions (respectively) "On Swampkinds," "Doubtful Intuitions," and "Absent Qualia" in *Mind and Language* 11, 1 (1996). Papineau argues that his teleological theory of representation can be thought of as reducing the everyday notions but that "the teleological theory can conform to everyday thinking in most cases without conforming in all. Like other worthwhile reductions, it may in some cases rectify naïve common-sense judgements [. . .]. In particular, it may rectify the everyday intuition that Swampman shares the contentful states of his normal doppelganger." Papineau, "Doubtful Intuitions," 132.

13. It does not follow, of course, from the fact that something is not a person that it is not apt for moral concern: consider animals. But surely this is far too little to bestow upon creatures that are behaviorally indistinguishable from us.

14. See Millikan, "On Swampkinds," and Karen Neander, "Swampman meets Swampcow." Dennett also attempts this defense in "Cow-sharks, Magnets, and Swampman," in *Mind and Language* 11, 1 (1996).

not really so.[15] Nor can it be considered a refinement of our ordinary, common-sense mental vocabulary (justified on explanatory grounds) to be told that although Swampman John may be acting *as if* he has a hatred of pigeons and strong convictions about the deleterious effects of their droppings, he does not really have either of these because he did not evolve.[16] Why should we accept as-if beliefs to explain real events? Or—contraposing—why should we demote what are indistinguishable from genuine actions to make-believe ones? Surely we cannot be expected to change our concepts whenever we encounter non-naturally-possible (e.g., purely fictional) contexts—or how else should we be able to make the necessary distinctions within them? We still need, for example, to be able compare Dorothy and the Scarecrow (who are both arguably persons although one is a human being and the other is made of straw) and contrast them with Toto who is a dog—just as we would want to contrast Glinda, the Good Witch, with the Wicked Witch of the West. The point is that the terms "person," "good," and "wicked" no more change their meaning because they are used as descriptions in fictional contexts than do the words "human being," "straw," or "dog." So why insist this must be so for propositional attitude and consciousness concepts?

To be sure, there may be real differences in our attitudes toward Swampman and human John. Swampman John, like human John, will square his shoulders and puff out his chest (just a little bit) when he talks about his days as a minor county cricketer. But whereas we have reason to believe most of what human John says, we might not be sure how to regard the chattering of Swampman John. Perhaps—after hearing a detailed story about that day in Gravesend when he (a young 17-year-old, batting against a pro who played for the county and for England) hit three boundaries in an over, and was cheered off the field after the pro decided to bowl a bouncer (remember there were no helmets in those days), which he tried to hook but played too early so that the ball looped as an easy catch to mid-on—perhaps we ought to

15. See Papineau, "Doubtful Intuitions"; Millikan, ibid., and Dretske, "Absent Qualia."
16. See Neander, "Swampman Meets Swampcow," 127.

consider making a distinction between the real (short) history of the life of a mere organism and the fictitious (long) history of the life of a real person.[17]

In accepting the invitation to delimit the application of our mental concepts in the light of the "theoretical advances" claimed by certain positions in contemporary philosophy of mind, we would be facing a dilemma. After all, if swamp-beings were indistinguishable from us except for their history, we would presumably still need a justified way of assigning and withholding blame. A usual factor in distinguishing someone who should be held responsible for his behavior from one who should not is to ascertain his intentions. If we were not allowed either to construe the behavior as actions, or attribute intentions, or assign responsibility, then either we would be forced to make up new concepts allowing us to make exactly the same distinctions (except that these new concepts would be correctly attributable to unevolved creatures) or we should be obliged to forego the rich descriptions that give any interest to the notion of "behavioral indistinguishability." But why should we agree to make up new concepts when they are doing exactly the same (explanatory) work as our old ones? The alternative suggestion (Papineau's)—that we should allow our "naïve, common-sense judgements" to be "rectified" and that we should learn to view the behavior of these creatures as mere noise and reflex—not only flies in the face of the phenomenon of conceptual change in science; it is morally unacceptable.[18]

These arguments do not depend on the thesis that sentences containing mental predicates are translatable without loss of meaning into sentences not containing them nor do they depend on the verification-

17. See Joe Levine's suggestion that Swampman's discourse be considered a form of fictional discourse in "SwampJoe: Mind or Simulation?," in *Mind and Language* 11, 1 (1996): 88.

18. Philip K. Dick's 1968 science fiction story *Do Androids Dream of Electric Sheep?* explores the moral implications of a community's persecution of (by and large) behaviorally indistinguishable androids. This story—upon which the film *Blade Runner* was based—is, fundamentally, a story about racism. The baselessness of the community's fear of androids is explored through the contrast with their passion and obsession for pets (electric or natural) and through the hunter-protagonist's growing empathy with his victims. The theme is also explored in Steven Spielberg's film *AI.*

ism that fueled logical behaviorism. Nor do these arguments depend upon the weaker thesis that sentences containing mental predicates can be translated without loss of meaning into sentences containing behavioral predicates and other mental ones. Nor do they imply the still weaker thesis that there are logical entailments between mental concepts and behavioral ones: each of the ascriptions to zombie John of the consciousness concepts considered above, while justifiable along the lines suggested, would also be defeasible in familiar ways (e.g., if zombie John were to pour the wine on the table instead of in the glasses, we would start to wonder about his vision). This leaves open the logical possibility that the lack of an inner, unknowable something may also serve as a defeating condition. But this logical possibility was basically accepted at the outset when, for the sake of argument, concept revision was not ruled out in principle. In the course of arguing that the particular concept revision that would permit the possibility of zombies and other mindless, but "behaviorally indistinguishable" creatures should be ruled out, my arguments have merely relied on the incontrovertible fact that explanations and justifications for the application of mental concepts tend to mention the actions (not to mention the beliefs, facial expressions, etc., and the circumstances) of the individual concerned. The argument strategy has been to show the distortion—with the loss of descriptive and explanatory power—that would result if our mental concepts were stretched in the ways envisaged.[19]

In order to remind ourselves of the meaning of mental concepts—those used in everyday, pretheoretical discourse—we need to keep

---

19. What is called "logical" or "analytic" behaviorism discussed in contemporary, introductory texts in philosophy of mind hails from the logical positivism of Carnap and Hempel. See, for example, Jaegwon Kim, *The Philosophy of Mind* (Boulder, CO: Westview Press, 1996). A much weaker thesis—closer to the later Wittgenstein and to Ryle and one that I have tried to elucidate in the text—is that the meaning of mental predicates is tied up with their natural expressions in speech and in behavior. Arguments of the kind I have assembled here would rule out not only zombies, but it would also suggest a problem with the X-worlders (or "Super-super-Spartans") of Hilary Putnam's "Brains and Behavior," (1965), in *Readings in Philosophy of Psychology,* vol. 1, Ned Block, ed. (Cambridge, MA: Harvard University Press, 1980). See chapter 8, this volume, "Pain, Polio, and Pride: Some Reflections on 'Becausal' Explanations."

sight of the justifications and explanations that we require and accept for their application. To suggest that the meaning of an expression is tied to the justifications for its use is not to imply that we will be called upon to justify its application on any particular occasion; still less is this discussion meant to imply that we learn of another's state of mind by inferring it from some outward sign. The fact that there are justifications and explanations that can be given for the application of a term is a simple consequence of the fact that the term is meaningful. Justifications and explanations for the ascription of sensation concepts are rarely required once an individual has mastered a language; this is especially the case for first-person ascriptions. It might help, therefore, to think of how these concepts are taught to children or to imagine circumstances in which there may be some doubt as to why they have been applied. Reflection on the occasions when explanations or justification for their use is appropriate will enable us to see that such considerations permit the ascription of mental concepts to creatures regardless of how they evolved or of what is or is not inside them. It is this that allows us to say, counterfactually, that if there were unnatural beings, let alone perfectly natural beings (like animals) that behaved enough like us then these creatures would be minded. The interesting philosophical and moral issues arise when we attempt to spell out what would count as behaving enough like us. But about creatures that are our behavioral duplicates under all possible tests there should be no question. Why should this practice change?

That the meaning of our mental concepts permits their ascription to creatures regardless of whether they have an inner theatre accessible with the mind's eye is an embarrassment for the Cartesian dualist or the friend of private mental objects. That it permits their ascription to creatures regardless of whether they have brains is an embarrassment for the type-physicalist. And that it permits their ascription whether or not the contents of their internal states are a product of evolution is an embarrassment for teleosemanticists.

Hilary Putnam pointed out the implicit racism behind the type-identity theory in the 1960s, Daniel Dennett has made the charge more recently against the defenders of zombies, and Louise Antony has also raised a concern about the moral implications of (causal-) historical theories of representation.[20] Thus it might be thought that (nonhistorical) functionalism—a view defended at times by at least two of these writers—is the only viable alternative. But the fact is that the meanings of our mental concepts—which are intimately tied to the justifications and explanations given for their application—also permit their ascription to creatures regardless of whether they have internal states that play a complex causal role. Although I agree with these writers that the very rich notion of behavioral indistinguishability in play in these philosophical thought experiments is more than enough to secure the right to apply the mental concepts unproblematically, it is another question whether functionalism can take comfort from this fact.

In the flight from dualism, most contemporary philosophers went along with the idea that our mental concepts pick out something natural: the obvious candidates to replace psychical or spiritual states were either physical states or physically realized functional states. But I am not convinced that the predominant role of mental concepts in organizing, categorizing, or explaining an individual's actions in the circumstances is served even by their purporting to refer.[21] Mental concepts seem to function by and large to "collect" powers or abilities, but there is no reason to construe these in turn as inner states that play a causal role.[22] To be sure, brain states are implicated in human beings (and other animals) in the exercise of these powers or abilities. They are,

20. See Putnam, "Psychological Predicates" (1967), reprinted as "The Nature of Mental States," in *Readings in Philosophy of Psychology,* vol. 1, Block, ed., 223–231; Dennett, *Consciousness Explained,* 405–406; and Antony "Equal Rights for Swamp-persons," in *Mind and Language* 11, 1 (1996): 70–75.

21. For a relatively rare acknowledgement of this position in the contemporary literature, see Georges Rey's discussion of what he calls "irreferentialism" in *Contemporary Philosophy of Mind—A Contentiously Classical Approach* (Cambridge, MA, and Oxford: Blackwell, 1997).

22. See Ryle's remark: "The explanation [of a person's action by a description of the workings of his mind] is not of the type "the glass broke because a stone hit it," but more nearly of the different type, "the glass broke when the stone hit it, because it was brittle." Gilbert Ryle, *The Concept of Mind* (1949) (London and New York: Routledge, 2009), 38. For a partial defense of this kind of position against post-Rylean critics, see part 2 (chapters 5–8) in this volume and my "Rethinking Ryle: A Critical Introduction to *The Concept of Mind,*" in Gilbert Ryle, *The Concept of Mind* (London: Routledge, 2009).

no doubt, part of the structure that makes the exercise of these powers physically possible. But there is no reason to identify (or functionally reduce) these powers with (to) these states any more than there is reason to identify (or functionally reduce) my ability to make you laugh with (to) an internal state of my body.[23] In fact I hope to have shown that we have good reason not to: refraining from making this identification would allow us to grant mentality and moral status to naturally impossible creatures who behave, per mirabile, similarly to the way that we do. Assurances that they will continue to behave (enough) like us under all possible tests should be more than enough to secure their mentality; the fact that there is no explanation of how such creatures could have these powers should make no difference to how our mental concepts are ascribed.

Of course we would be tempted to withdraw certain mental predicates upon discovering that a creature in the *actual* world is hollow, merely mechanical, or made of straw. But this is because, in the world as we know it, such a creature will not be able to sustain behavior that is "enough like us" under any reasonable tests let alone under all possible ones. But it is the conceptual, not the natural, possibility we are considering when we think about zombies, swamp-beings, and other "behaviorally indistinguishable" creatures; these creatures are and will continue to be, as a matter of stipulation, our behavioral twins. In introducing a creature as complex as John, I am hoping the reader will resist the invitation either to change or alter the predicates that we need to understand him, even if it were "discovered"—counter to the laws of nature—that he has not evolved or that he is made of straw. If anything could count as such a discovery, then it should be our science that is revised and not the use we make of mental concepts.

Those interested in defending a scientific study of the mind will no doubt continue to bite the bullet in the face of arguments that point to the unappealing consequences of their claims. As in all scientific

23. See Gordon Baker and Peter Hacker's discussion of vehicle-reductionism in *Wittgenstein: Meaning and Understanding* (Chicago: University of Chicago Press and Oxford: Blackwell, 1980), chapter 16.

investigations, they argue, new criteria are being proposed for the application of the relevant concepts, and the proposed change is justified on explanatory grounds. I hope to have shown, however, that there are overriding reasons, also broadly "explanatory," for rejecting these proposals. The acceptance of these new criteria would not merely require a willingness to deviate from common-sense psychology. Having indicated the moral consequences that arise when "language goes on holiday," I have suggested that their acceptance might present a threat to common decency as well.

CHAPTER ELEVEN

Conceptual Analysis, Theory Construction, and Philosophical Elucidation in the Philosophy of Mind

1. In 1983, Barry Stroud wrote that Wittgenstein's reputation was largely in eclipse. This was not, Stroud said, because the major themes of his work had been appreciated and absorbed into a tradition now busily engaged in extending them in new directions. On the contrary: his major themes had been either rejected or ignored.

> The more empirical, "naturalistic" turn in the approach of many contemporary philosophers, their search for "theories" and their appeal to general "theoretical" considerations apparently continuous with natural science . . . puts [contemporary] philosophy . . . farther from the spirit as well as the letter of Wittgenstein's conception of philosophical problems. He thought that "philosophers constantly see the method of science before their eyes, and are irresistibly tempted to ask and answer questions in the way science does. This tendency is the real source of metaphysics, and leads the philosopher into complete darkness."[1] A report on the main

1. The original source of the quotation is Ludwig Wittgenstein, *The Blue and Brown Books* (1958) (Oxford: Blackwell, 1975), 18.

> activity in the philosophy of mind in the period since [the mid-sixties when Wittgensteinian influences could still be discerned] would therefore be a report of activity within what Wittgenstein would regard as "darkness"; it would not be a report of developments and extensions of his own ideas during that period.[2]

Sixty years after the publication of the *Philosophical Investigations* contemporary philosophy of mind seems even farther from the spirit of the later Wittgenstein than when Stroud wrote this some thirty years ago. For philosophers are even more inclined to see the method of science before their eyes, and are even more irresistibly tempted to ask and answer questions in the way that science does. Georges Rey, for example, is quite explicit about this and (using a peculiar contrast to make the point) announces at the outset that his treatment of the mind "aspires to be continuous with science, not literature. . . ."[3] He explains that there is a recent resurgence of interest in the philosophy of mind with "explanatory questions" about what sort of thing a pain, a thought, a mental image, a desire, or an emotion is. Traditional philosophical theories of the mind are inadequate since neither materialism nor dualism provides a "serious" theory about the mind, which will give us a "serious" explanation of mental phenomena. And, according to Rey, though Wittgensteinian grammatical investigations may have given us a "heightened sensitivity to complexities and nuances of our ordinary mental talk," they "tended to occur at the expense of further theorizing about the mental phenomena themselves."[4]

As Rey acknowledges later in the book, however, it is not simply that such grammatical investigations *tended* to occur at the expense of

2. Barry Stroud, "Wittgenstein's Philosophy of Mind," *Contemporary Philosophy: A New Survey, Vol. 4: Philosophy of Mind,* Guttorm Florstad, ed. (The Hague: Martinus Nijhoff, 1983), 320.

3. Georges Rey, *Contemporary Philosophy of Mind—A Contentiously Classical Approach* (Cambridge, MA, and Oxford: Blackwell, 1997), xiii. Note, incidentally, that the idioms of "ordinary" mental discourse are not, of course, restricted to novelists. They are also those of doctors, lawyers, judges, teachers, priests, parents, friends, and even scientists: indeed everyone who wields the concepts of reason.

4. Ibid., 4.

further theorizing about the nature of mental phenomena: they were *designed* to do so. The "heightened sensitivity" attained by these investigations was to reveal that traditional philosophical conundrums (like the mind-body problem) result from being misled by superficial grammatical similarities between what are, as a deeper analysis will reveal, different kinds of expressions. Wittgenstein urged that the way forward is not to theorize about mental phenomena per se, but rather to investigate the various functions of expressions deploying mental concepts. But this stance, and the kind of investigation recommended by it, has led many contemporary philosophers to reject Wittgenstein's attitude as simply prejudiced against scientific enquiry. Rey parodies this attitude—crystallized in Wittgenstein's comment that explanations have to come to an end somewhere—by suggesting that

> one could . . . [rationally] . . . never ask for explanations of anything at all: we *could* just say that it is a natural capacity of lightning to burn what it strikes. The question is whether we *can also* rationally ask for slightly deeper explanations, and, if we can, what those explanations might be.[5]

This dismissal is too quick. Wittgenstein does not merely assert that explanations must come to an end. The *Philosophical Investigations*—and especially Wittgenstein's discussion of rules—presents a sustained argument for this claim. Nor does Wittgenstein simply reject scientific theories that are advanced about the phenomena in question: he disputes the philosophical presupposition that funds the scientific program—namely that expressions containing mental concepts always function to pick out independent states of affairs (objects, events, states, properties, or relations) that are amenable to a scientific investigation in the first place. This attitude is not antiscience: it is a reaction against a too-crude philosophical theory of meaning.

The recoil against this theory of meaning should not be exaggerated. We may agree that the uses of the word "brain" and its cognates can be

5. Ibid., 5.

regimented in such a way as to make it plausible that (on its primary use) it refers to something—an organ—whose nature it is the business of neurophysiologists to discover. It is another matter whether mental concepts, which are arguably abstract, should be so understood.

There seems no point in legislating for or against this question in advance of detailed argument, either via an appeal to the authority of what Wittgenstein has (allegedly) shown or in favor of science. Conceptual revision that is recommended by the advances of science will have a point when the concepts in question contain inconsistencies that present difficulties, or simply when they need to be made more precise for theoretical purposes. Deciding the matter will involve reinvestigating the problems that arise when we assimilate what may indeed be different kinds of expressions, as well as different kinds of explanations, and reminding ourselves what we need the relevant concepts for.

Since the late 1960s, many philosophers have by and large succeeded in placing the methods of the cognitive sciences and their philosophical *functionalist* and *mental representationalist* cousins at center stage in discussions about the nature of the mental. In so doing, they have wrested the subject from the grasp both of analytic, a priori philosophy and also from that of the philosopher who sees her work as continuous with science: reductive physicalists or eliminativists who do not want to accord the mind as prominent a role as the "mentalists" wish to give it. The result, however, is a strange sort of hybrid: a subdiscipline with many points in common between traditional analytic philosophy and the more empirically oriented work that is prevalent today.

Nonetheless, an interesting movement is worth noting. Philosophers are questioning cognitive science's close relationship with classical analytic philosophy.[6] As this occurs, some of Wittgenstein's arguments

---

6. Here I shall discuss Stephen Stich, "What Is a Theory of Mental Representation?," (1992), in *Mental Representation,* Stephen Stich and Ted Warfield, eds. (Oxford: Blackwell, 1994), and Michael Tye, "Naturalism and the Mental," *Mind* 101 (1992): 421–441. Per Sandin, "Has Psychology Debunked Conceptual Analysis?," *Metaphilosophy* 37, 1 (2006): 26–33 comes to conceptual analysis's defense and cites Michael Bishop, "The Possibility of Conceptual Clarity in Philosophy," *American Philosophical Quarterly* 29, 3 (1992): 267–277; William Ramsey, "Prototypes and Conceptual Analysis," *Topoi* 11 (1992): 59–70; and

against analysis are reemerging. This will become evident by reviewing the critical literature on a project that occupied many philosophers of mind in the mid-1980s and 1990s: the quest for (the semantics for) a naturalistic theory of content or a theory of mental representation.[7]

In section 3, I consider one rationale for constructing such a theory: to uncover the (alleged) store of tacit knowledge delivered in the (alleged) mechanisms underpinning our ability to understand and interpret one another using mental concepts. In section 4, I consider a different rationale: one which claims to eschew philosophers' "armchair" intuitions altogether. I argue that neither project can proceed in a way that is independent of our considered judgments about the correct employment of ordinary mental concepts. En route, I show that Wittgenstein's arguments not only cast doubt on classical conceptual analysis: they throw into turmoil the entire philosophical underpinning of the cognitivist approach.

Before looking at the rationales for constructing theories of content, it will be useful to clarify the meaning of the terms "content," "propositional attitude," and "mental representation," which are used variously in the contemporary literature. These are technical philosophical terms and their relation to expressions such as "belief," "desire," "thought," etc. is not always clear. To illustrate this difficulty, I shall trace a well-traveled pathway through the contemporary literature: one that culminates in a call for a theory of mental representation.

2. It is a grammatical fact, as philosophers used to say, that many of our common, ordinary, everyday expressions containing mental verbs ("want," "think," "wish," "believe," "judge," "guess," etc.) often

---

Michael DePaul and William Ramsey, eds. *Rethinking Intuition: The Psychology of Intuition and Its Role in Philosophical Inquiry* (Lanham: Rowman & Littlefield, 1998) as others who have challenged the relationship. See also Michael Bishop and J. D. Trout, "The Pathologies of Standard Analytic Epistemology," *Nous* 39, 4 (2005): 696–714, who charge that standard analytic *epistemology* accepts constraints that depend on the "considered epistemic judgments of a small group of idiosyncratic people" (ibid., 697).

7. The project may have moved offstage, but any position that has invested in the theoretical notion of mental representation or mental content is in effect betting that it will succeed.

have as their subject a pronoun or proper noun and are followed by an accusative noun "that . . ." clause. It is also a fact that such expressions are often conjoined with the connective "because" to answer "Why?" questions, such as "Why did she open the door? She opened the door because she wanted to let the cat in." These grammatical facts, which are tedious to spell out, are often shortened or abbreviated by saying that we ascribe (what Russell called) "propositional attitudes" to a person in order to explain her behavior.

Now the so-called *realism* upon which theories of mental representation are based begins with the idea that such expressions, as used in such contexts, function in a particular way: to name a kind of mental state, event, or disposition (also called a "propositional attitude") whose ontological nature is open to investigation. This much is supposed to be largely explicit in common sense.[8] Some philosophers also hold that it is implied in this so-called common-sense view that propositional attitudes have a dual nature: they consist of states or attitudes that are the referents of the mental verbs, and these attitudes or states "have," "take," or "are the bearers of" meaning or content—in a way to be investigated—where the content is the referent of the sentence complement that follows the (attitude) verb.[9] Most philosophers also believe (and argue that it is implied by this allegedly common-sense view) that the connective "because" (following the "Why?"-question) signals a causal explanation. This fuels the idea, in turn, that what is referred to in the propositional attitude expression must (at least in part) be or signal the occurrence of that which is widely considered necessary for being a cause: a logically independent, antecedently occurring (triggering) event, which enters into law-governed causal relations.[10]

---

8. See, for example, Rey, *Contemporary Philosophy of Mind,* xiii; Jaegwon Kim, *The Philosophy of Mind* (Boulder, CO: Westview Press, 1996), 125; Jerry Fodor, "Fodor's Guide to Mental Representation: The Intelligent Auntie's Vade-Mecum," (1985), in *Mental Representation,* Stich and Warfield, eds., 11.

9. See Rey, ibid., 147–151, and Michael Devitt and Kim Sterelny, *Language and Reality—An Introduction to Philosophy of Language,* 2nd edn. (Oxford: Blackwell, 1999), 137–138.

10. See Donald Davidson, *Essays on Actions and Events* (Oxford: Clarendon Press, 1980), 11, and Fodor, "Fodor's Guide to Mental Representation," 10. With the introduction of propositional attitudes conceived as mental states (as here understood), the move that

An investigation of the nature of mental states—especially propositional attitude states—has been one of the main focuses of philosophy of mind. Functionalism attempts to accommodate the abstract nature of mental states by supposing they can be implicitly defined all at once and in nonmental terms by a story (or "ramsification") that specifies a (potential) role in a causal network. According to some versions, the ramsification quantifies over first-order properties, or "realizations" of mental states, whose nature is left unspecified, which are the alleged occupants of these causal roles.[11] The story describing this causal structure implicitly defines these properties as the types of mental states they are. This causal-role functionalism thus attempts to accommodate the autonomy of the causal-explanatory role of the mental properties on the basis of their realizations' external or causal/nomological relations to other properties, and not on the basis of any (compositional or structural, intrinsic) details of the occupiers of the role. In principle anything (whether it be mechanical, electrochemical, or spiritual) that fills the causal role specified counts as having the second-order, functional or mental property. In practice, most functionalists think that mental states are realized by (first-order) internal, physical, or microphysical states of the organism.[12]

Wittgenstein was keen to question is in place: that expressions containing mental verbs taking "that"-clauses as complements and used to answer "Why?"-questions function as names, let alone as names of types of internal (first- or second-order) states of an individual or system that causally interact, and whose ontological nature is open to philosophical, and eventually, perhaps, empirical investigation. Note that Davidson's sophistication on this traditional story is that all events that fall under mental kinds also fall under physical kinds; but only insofar as the two events that causally interact fall under physical kinds are they subsumed by genuine laws.

11. See Jaegwon Kim, *Mind in a Physical World: An Essay on the Mind-Body Problem and Mental Causation* (Cambridge, MA: A Bradford Book, MIT Press, 1998/2000), and Ned Block, "Can the Mind Change the World?" *Meaning and Method,* George Boolos, ed. (Cambridge: Cambridge University Press, 1990), 137–170.

12. See Kim, ibid., 12–23, and Rey, *Contemporary Philosophy of Mind,* 176–180. This (causal-role) functionalist is sometimes contrasted with another (token-identity or "realizer") functionalist (e.g., Lewis) who identifies the referent of the mental state term with the (presumed-to-be) physical realization of the causal role.

Suppose this is so. What is the relation between causal-role functionalism and common-sense explanations? Originally, functionalism was proposed as a theory about the meaning of mental terms that are used every day in nontheoretical discourse.[13] The concept of pain, on this account, is thought to pick out an (unspecified) inner state of an organism (or system) that occupies a certain causal role in mediating between other inner mental states, input, and behavioral output. This causal role is to be specified by common-sense platitudes: e.g., that pain is likely to be caused by tissue damage and result in avoidance behavior. Another version suggests that the role is instead to be found by the traditional methods of a priori philosophical analysis.[14] This approach would, like the first, dovetail with the idea that mental concepts' primary domain is common-sense explanation but only if the methods of a priori philosophy are to make explicit what is already implicit in our use of mental concepts.

Psychofunctionalists, by contrast, have a more ambivalent attitude to ordinary mental concepts.[15] They agree with the others that mental concepts function to pick out a causal role realized by some or other inner (physical) state. They also agree that a complete, constitutive account of the second-order or functional states will be given by a story outlining the causal relations between their occupants and their relation to input, output, and other mental states. They believe, however, not only that the occupants of this role, but that the role itself is to be discovered by empirical psychology.[16]

We might wish to press them on the relation between what may turn out to be two different subject matters—common-sense psychol-

13. David Lewis, "Psychophysical and Theoretical Identifications," *Australasian Journal of Philosophy* 50, 3 (1972): 249–258.

14. Sydney Shoemaker, "Some Varieties of Functionalism," (1981), *Identity, Cause, and Mind* (Cambridge: Cambridge University Press, 1984).

15. Ned Block, "Troubles with Functionalism," (1978), *Readings in Philosophy of Psychology,* vol. 1, Block, ed. (Cambridge, MA: Harvard University Press, 1980), and Rey, *Contemporary Philosophy of Mind.*

16. Block, ibid., 268–305, and Rey, ibid., 187.

ogy and empirical psychology. *Psychofunctionalism* is prepared to be theoretically revisionist about mental concepts since, according to this doctrine, the material (e.g., descriptions) in which the concept figures, or which picks out the concept or its referent, are merely "reference fixers," not only of the alleged first-order inner states that occupy or realize second-order, functional roles, but also of these roles themselves. Thus psychofunctionalists are also prepared to be revisionist about the explanatory role of the common-sense answer to the "Why?"-questions that invoke mental concepts. The thought must be that common-sense psychological explanations (e.g., "she opened the door because she wanted to let the cat in") are also rough-and-ready approximations to the real causal/explanatory story that an advanced empirical psychology will one day deliver.[17]

On one use of this technical term, "mental representations" are, unlike propositional attitude or other mental (state) expressions, alleged *not* to belong to common sense. They belong, rather, to a philosophical theory that attempts to account for the propositional attitude states that *are* allegedly picked out by common-sense mental expressions.[18] A theory of mental representation is supposed to explain, among other things, how an individual could be in an inner state that is both content-bear-

17. Indeed, this seems one way of construing the position to which Davidson is committed given his adherence to both the anomalism of the mental and to the nomological character of causation. The "because" introducing a mental phrase (e.g., "she wanted to") given in answer to the "Why"-question signals a causal relation, but the relata are subsumed under laws only when picked out under their physical descriptions. Events, for Davidson, enter into causal relations no matter how they are described. But presumably it is only when we know these laws (if ever) that we will be given a causal explanation of the action. The problem however, that it was in virtue of its being thought necessary for explanation that led Davidson to construe reasons as causes in the first place. There is thus an equivocation, it seems, in Davidson's use of "explanation": in one case it implies understanding and in the other it does not. For more discussion, see "Why Reasons May Not Be Causes" (chapter 5).

18. Note that, strictly speaking, to hold that the mind "somehow" represents states of affairs is not yet to be committed to the existence of mental representations, and a commitment to the latter does not yet yield a commitment about their nature (e.g., pictorial or linguistic). In the text I follow the predominate view.

ing and causally efficacious. Those who posit mental representations, on this understanding, may be seen as augmenting a broadly functionalist framework by accepting that the states or attitudes are defined as a particular type of state (belief, hope, fear, etc.) by a story or computation of a certain kind (e.g., "by appealing to the familiar flow-charts of cognitive psychology"[19]). The most common view—in order to account for their productivity and systematicity (among other things)—is that propositional attitude states in particular involve relations to inner symbols. An individual's having a specific propositional attitude (now construed as her being in a causally efficacious state B that bears a specific content C) consists in her being in a B relation (to be specified by some story) to an inner symbol—or mental representation—that means that C.[20] For a propositional attitude expression correctly to describe an individual—say, "she wants to let the cat in"—that individual must be in the "want relation," usually specified along broadly functionalist lines, with an inner symbol meaning that her cat be let in.

The Language of Thought hypothesis is committed to a further supposition about the nature of the inner symbols.[21] It holds that they are sentences (rather than, say, pictures or mappings) belonging to a language of thought that are entokened in the brain—say, as electrical impulses. The computational theory of mind says that these sentences are manipulated in a rule-governed way: they are computed or processed in accordance with systematic formation and transformation rules, many of the relevant semantic properties of which are "mirrored" in syntactical ones. It is this computational process that is alleged to account for the systematicity, rationality, and consistency that is exhibited by many of our cognitive or thought processes.

---

19. Georges Rey, "Language of Thought," (1998), in *Routledge Encyclopedia of Philosophy,* Edward Craig, ed. (London: Routledge, 2004).

20. See Hartry Field, "Mental Representation," (1978), in *Readings in Philosophy of Psychology,* vol. 2, Block, ed. (Cambridge, MA: Harvard University Press, 1981); Jerry Fodor, *The Language of Thought* (New York: Crowell, 1975); and Fodor, "Fodor's Guide to Mental Representation."

21. Fodor, *The Language of Thought,* and Rey, *Contemporary Philosophy of Mind.*

is a constraint, accepted by some but not by all, that a theory of mental representation must show how what seems on the face of it to be a relation between an individual and some sort of abstract object can be accommodated without positing unreduced entities (Propositions or Thoughts) or epistemic relations (such as the "grasp" of one).[22] Language of Thought theorists may attempt to meet this constraint by supposing either that the syntax of the language determines its meaning (as a more robust, conceptual role, functionalist approach might do) or by proposing some separate sort of account (e.g., causal covariational or teleological) of the meaning of the sentences.[23] Providing such an account would be one point of a theory (of the semantics) of mental representation.

The broadly functionalist approach, and especially its computational version, is thought to offer a philosophical rationale for the underlying assumption in cognitive science: namely, that mental abilities such as thinking, deliberating, acting, and interpreting involve the systematic, internal processing of semantically valued symbols. This approach asks how the abilities to be explained can be turned into problem-solving tasks; it then tries to discover the rules, heuristics, and brute procedures for getting a device to produce correct responses. Classical cognitive science considers that such abilities can be construed as information problems to be solved by the step-by-step manipulation of symbol/representations according to a set of rules. Nonclassical approaches are also committed to the information-processing paradigm, but disagree about what is represented and about the processes used to solve the problem.[24] It is a further conviction, but one shared by many, that this sort of empirical work will in turn produce constitutive answers to

22. Fodor, "Fodor's Guide to Mental Representation."

23. For conceptual role, functionalist approach see Field, "Mental Representation," and Ned Block, "Advertisement for a Semantics for Psychology," (1986), in *Mental Representation,* Stich and Warfield, eds., 81–141. For the causal covariational version, see Jerry Fodor, *Psychosemantics—The Problem of Meaning in Philosophy of Mind* (Cambridge, MA: MIT Press, 1987). For the teleological approach see Karen Neander, "Misrepresenting and Malfunctioning," *Philosophical Studies* 79 (1995): 109–141.

24. Andy Clark, *Microcognition: Philosophy, Cognitive Science, and Parallel Distributed Processing* (Cambridge, MA, and London: A Bradford Book, MIT Press, 1989).

questions about the nature of these mental abilities.[25] This, however, is highly debatable. Or so I shall argue.

I have marked but one fragment of the territory in philosophy of mind in order to pose a general question about the relation between mental representations, as they are understood on one use of the term, and the concepts we typically use to describe individuals who are minded. The particular strand I have summarized here—which is considered the orthodoxy insofar as others define and refine their own views against it as the backdrop[26]—is motivated at the outset by our nonphilosophical and nontheoretical use of mental concepts, especially propositional attitude expressions. But though propositional attitude *states* are acknowledged to have close ties with common uses of propositional attitude *expressions* (they are the alleged referents of such expressions) mental representations are sometimes thought to have no *direct* links with these expressions. They are often explicitly characterized as theoretical posits whose relation is at a considerable remove from ordinary mental concepts and are hence impervious to criticism that issues in reminders about how we use mental idioms.

The literature does not always distinguish, however, the differences between the propositional attitude expressions we apply to individuals, the alleged referents of these expressions, which I have called "propositional attitude *states*," and the theoretical posits called "mental representations," construed as inner symbols, that form part of a theory of what it is for a system to have or be in these states. Both "mental content" and "mental representation," as we shall see, are often used to mean either

25. Stich cites William Lycan, *Judgment and Justification* (Cambridge: Cambridge University Press, 1988), as one philosopher who is prepared to admit that mature psychology will reveal the true nature of the mental phenomena at which our ordinary mental concepts merely gesture. Rey, too, is prepared to be revisionist about ordinary mental concepts. For a thoughtful discussion see Clark, ibid., 54–58.

26. Ned Block, "The Mind as the Software of the Brain," in *An Invitation to Cognitive Science,* vol. 3, D. Osherson, L. Gleitman, S. Kosslyn, E. Smith, and S. Sternberg, eds. (Cambridge, MA: MIT Press, 1995).

the states to which our expressions are alleged to refer or to the theoretical posits (inner symbols) of a theory designed to explain these states.

In the following sections, I shall explore the rationale for constructing a theory of mental representation on both understandings of this notion. I shall argue that both projects are subject to reminders about the correct employment(s) of mental terms. Those who would reject the appeal to linguistic use are mixing up conceptual analysis as traditionally conceived with the cartographical or topographical elucidation of concepts recommended by Ryle and Wittgenstein.

3. Stephen Stich, noting that there are different reasons for constructing a theory of content in the philosophical literature, has rightly urged that we consider in detail what such a theory of content or mental representation is supposed to do. He characterizes one motivation as follows:

> A prominent feature of our everyday discourse about ourselves and about other people is our practice of identifying mental states by adverting to their content . . . [Here Stich mentions the grammatical facts; the philosophically crucial move of positing propositional attitude states as the referents of our expressions is never questioned.] In these, and in a vast range of other cases, the attribution of content is effortless, unproblematic, and unquestionably useful. Moreover, in the typical case, there is widespread intersubjective agreement about these attributions. Plainly, there must be a mental mechanism of some complexity underlying this ubiquitous practice, and it seems plausible to suppose that the mechanism in question includes a store of largely tacit knowledge about the conditions under which it is (and is not) appropriate to characterize a mental state as the belief or the desire *that p*.[27]

Stich suggests that a goal for a theory of mental content might be to describe the body of tacit knowledge that is stored in the mechanism

27. Stich, "What Is a Theory of Mental Representation?," 350.

underlying our quotidian practice. A theory of mental content, on this conception, is a theory of propositional attitude *states,* which should specify what is known by the ordinary person enabling her to ascribe propositional attitudes and also (presumably) enabling her to explain (understand, communicate with, manipulate, anticipate, etc.) the one she is interpreting.

If a theory of content, understood in this sense, is to explain how a participant in our interpretive practices comes to an understanding of her subject by ascribing or "identifying" mental states then the theory is answerable to the kind of understanding that is involved, and thus the kind of explanation that is delivered, in such nonphilosophical and nontheoretical contexts. Even if a theorist studying these practices is the best person to articulate what constitutes their rules or suppositions, it is still the ordinary practices, explanations, and understanding they deliver that constitute the target of the investigation. On this construal of the role of a theory of content, the relation between the theory and the common-sense knowledge that is accessible from our armchairs is clear and direct.[28]

Philosophers working on theories of mental content, however, have a very ambivalent attitude toward ordinary talk. "Empirical philosophers" make it clear that the authority on the nature of the allegedly causal-explanatory role as well as the states that are alleged to enter into causal relations is within the province of cognitive science whose theorists should not feel constrained by the remarks of philosophers interested in examining the use of mental expressions. Those tempted to appeal to ordinary language are often accused of interfering with a perfectly legitimate scientific enquiry.[29] This issue is obscured by the

---

28. The armchair-bound philosopher is reflecting on the practices of her linguistic community. Such a task is neither prior nor posterior to "experience" in one of the senses intended for the distinction: namely, when *"a priori"* connotes the grasp of a truth via the intuition of the faculty of reason.

29. Robert Cummins, *Meaning and Mental Representation* (Cambridge, MA and London: A Bradford Book, MIT Press, 1989), chapter 1 is an example; he is explicit that his discussion of mental representation is about the theoretical posits presupposed within (the various frameworks of) cognitive science. See also note 25, above.

various senses of "mental content" and "mental representation." As the above reflections show, even if this stance could be maintained for second-order theoretical posits (which, in any case, I shall go on to deny) it is impossible to see how it could be maintained for first-order ones when the explanandum is the (ordinary) application of (ordinary) mental concepts and the (ordinary) understanding it delivers.

It is worth pausing for a moment to wonder what accounts for this uneasiness with the appeal to ordinary language. Why, when mental concepts have such an entrenched role in our lives, good enough for general purposes, should we be tempted by the view that the deliverances of science may radically alter it? One explanation for the uneasiness lies with the background assumptions of the classical analytical approach. As Stich and Tye point out, that these assumptions have been absorbed is evident in the way that theories of mental representation are constructed and debated.

In order to understand this point, it will be useful to recall G. E. Moore's explicit characterization of analysis—as the clarification of concepts—which says that the verbal expression of analysis must follow a standard pattern of paraphrase in which what is analyzed is logically equivalent to a larger, more explicit, and synonymous expression. Those motivated by scientific realism, and *psychofunctionalists* in particular, do not suppose that their theories must be constrained by the meaning of ordinary concepts in this tight sense, hence the demand for synonymy of expressions is no longer in place. Instead, as I have indicated, the strategy might be to suppose that ordinary language contains the material to fix the reference of what can be discovered a posteriori to be some property or set of properties. On this conception, any proposed identity claim linking the concept with the alleged referent will purport not to be definitional, but synthetic. This in its turn may invite a reforming definition. Nonetheless, any theoretical or reforming definition might still be subject to the challenge that it has either left out something important, or that it has added something inessential.[30]

30. For a discussion of attempts to provide such definitions in ethics and the relation to Moore's "open question" argument, see Connie Rosati, "Naturalism, Normativity, and

Thus in deeming that a theory of propositional attitude (states) should specify necessary and sufficient conditions for a mental state *M* to have content *C*, one is following closely in Moore's footsteps. In suggesting that these conditions should be specified *naturalistically*—or without mentioning semantic or intentional notions—one is still following closely in Moore's footsteps (though Moore would presumably be as suspicious about naturalism in this domain as he was elsewhere).

A number of philosophers have questioned the viability of this project.[31] Stephen Stich and Michael Tye have doubts partly on inductive grounds: since conceptual analysis or conceptual definition as Moore conceived it has, by and large, been a failure, then so too will conceptual analysis with even tighter constraints about what is allowed in the paraphrase.[32] The fact that we are really operating with old-fashioned conceptual analysis, disguised as theory construction, is evident from the fact that the putative theorems (of the form X believes *P* if and only if . . . , or mental state *M* has content *P* if and only if . . .) are rejected when a counterexample can be found in which ordinary intuitions tell us that X believes *P* or mental state *M* has content *P* even though the putative condition does not hold (or vice versa). Even if the theory is intended as a reforming one (in which the theorems are intended as theoretical identities rather than as analytic definitions), it is still accountable to the charge that something important has been left out or unnecessarily added. The proposed revision is thus acceptable or dismissible in part on the basis of how important this omission or addition is thought to be.

The past failures of conceptual analysis are not the only grounds for pessimism, however. Both Stich and Tye have also claimed that empirical findings support rival theories about the nature of the mental structures that underlie people's judgments when they classify items into categories. The empirical evidence shows that these structures do not exploit tacitly known necessary and sufficient conditions. Both

---

the Open Question Argument," *Noûs* 29, 1 (1995): 46–70.

31. See note 6, above.

32. Stich, "What Is a Theory of Mental Representation?," and Tye, "Naturalism and the Mental."

philosophers cite Eleanor Rosch's work on prototypes as an example of such a rival theory, and both mention in a footnote her indebtedness to Wittgenstein's work in the *Philosophical Investigations.*

What conclusion do these philosophers draw from this empirical evidence? Stich speculatively concludes that the ability to categorize intentional concepts such as *believing that P,* or *desiring that Q,* will not be explained by, nor have underlying them, mental mechanisms that utilize classical concepts of the sort that can be defined by sets of necessary and sufficient conditions. If our intuitions about whether a state has the content *that P* are in part *constructed,* and not *determined* by rules for the application of the concept but are rather guided in part in response to the circumstances in which the judgment is called for, then there are no such rules in the form of necessary and sufficient conditions for a theory of mental representation to deliver. Stich concludes that

> . . . if using the method of definition and counterexample is the hallmark of a philosophical theory in this area, and if the commonsense concept of mental representation is like every other concept that has been studied empirically, there is a sense in which *there can be no philosophical theory of content.*[33]

To a student of Wittgenstein (ignoring, for the moment, the reference to mental mechanisms), all of this is well and good. But Stich goes on to suggest that if this is what a theory of mental content is for, the theorist will probably have to give up doing linguistic philosophy and start doing cognitive science instead.[34]

This is an extraordinary claim. For how can cognitive science help? If Stich and Tye are right to doubt that an analysis will be discovered offering sufficient and/or necessary conditions for something, say, to be a state with content we should—for similar reasons—also be suspicious of the idea that our ability to apply mental concepts has a body of

33. Stich, "What Is a Theory of Mental Representation?," 354.
34. Ibid., 354.

largely tacit knowledge that underlies it. For what is the content of this knowledge if it is not the sort of theory Stich has rejected?

The difficulty can be seen by asking how the knowledge is supposed to be delivered. The usual supposition is that an interpreter is "caused" to deliver correct interpretations most of the time because the rules of interpretation are embodied in mental mechanisms. This presumably means that having the body of knowledge is causally sufficient to produce the correct answers, provided that the mechanism is not defective. But if the mechanism that embodies the knowledge is causally sufficient to produce the right answers, then the content of knowledge itself, as it were, would have to suffice. This leads us back to the idea that the knowledge consists in at least *sufficient* conditions for the application of our content concepts. The requirement that the knowledge consist in *necessary* conditions would ensure that the principles or theorems are actually consulted or processed in delivering it. But surely the rejection of a classical conception of concepts brings with it the rejection of the model of explanation that takes the content of knowledge to be sufficient and/or necessary to determine correct interpretations. To put the point another way, surely the rejection of the idea that there are rules that determine the correct application of mental concepts entails the rejection of the idea that our ability to interpret is a matter of knowing these rules. What room is left, then, for the idea that the ability to be explained amounts to a kind of information-processing?[35]

The worry, developed by Ryle and Wittgenstein, is that the mechanism hypothesis is misguided. We may defend the claim that an individual has applied mental concepts by showing how her use accords with the meaning or the norms, standards, or principles that govern the practices in which we use the relevant expressions. Although sometimes a learner of this practice can be taught some of its moves by appeal to such meanings, she is not required to follow them or their expressions

35. The cognitive scientist retreats to a weaker understanding of tacit knowledge at her own risk: if the alleged representations or the rules that supposedly govern their operations are only "read off" from normative practices by a theorist (or programmer) then information-processing has disintegrated into mere know-how. This is discussed further in chapter 12, "Ryle's Regress and the Philosophy of Cognitive Science."

in the more robust way demanded by the cognitivist paradigm. It may make sense to explain certain abilities by showing how an individual uses expressions of norms to guide her, but this is not always necessary, nor, when it comes to certain abilities, is it even coherent.[36]

Stich and Tye are right that analytic philosophy, with its commitment to classical concepts and its attendant epistemological problems, influences sophisticated attempts to construct theories of content, construed in the first sense as the store of knowledge that grounds our interpretive abilities. But they fail to see that these same commitments and problems infect the entire philosophical underpinnings of the cognitivist approach.

Stich and Tye are also wrong to suggest that there is no place for armchair philosophizing. For even if cognitivism could somehow survive this criticism, it would still be misleading to suggest that it could discover the solution to the difficult questions that a theory of content would have, eventually, to face: namely, to say what *counts* as someone's having a particular mental attitude with a specific content, or what counts as an appropriate deployment of the relevant mental concepts. To put the point another way, experimenters testing underlying mechanisms for their subjects' ability to "identify these states" have to take a stand on whether any particular ascription of content is successful before any (ostensible) mechanism for delivering these (and only these) successful judgments can be investigated. To understand our interpretive abilities we need to pay close attention to how we justify, defend, and correct these attributions. This is precisely what Wittgenstein, Ryle, Austin, Anscombe, Melden, and other so-called "ordinary language" philosophers set out to do: describe, systematize, and make recommendations for the (correct) employment of mental predicates. These philosophers,

36. The problem, as Wittgenstein and Ryle argued, is that rule-following is itself an activity that requires certain broadly rational abilities. Any rule-following explanation, or an explanation in terms of prior theoretical operations, will presuppose, and therefore cannot explain or account for, these broadly rational abilities. I develop this further in chapter 3, "Playing the Rule-Following Game," and chapter 4, "Real Rules." See Ludwig Wittgenstein, *Philosophical Investigations,* G. E. M. Anscombe, trans. (Oxford: Blackwell, 1953), and Gilbert Ryle, *The Concept of Mind* (1949) (London and New York: Routledge, 2009).

too, rejected the underlying theory of meaning grounding classical analytic philosophy; thus their project is not subject to the same criticisms that face classical (definitional), a priori conceptual analysis.

4. I HAVE EXPLORED one reason for constructing a theory of mental representation or content: to articulate the knowledge that embodies the ability of ordinary folk to ascribe mental concepts to themselves and others. The supposition was that this knowledge is delivered in some underlying mechanism. I have argued that there is reason to doubt the supposition, but even if it were correct, the project clearly cannot proceed in a way that is independent of our ordinary practices (and thus of our considered judgments) about when it is correct or not to "attribute content"—ascribe mental concepts—because this was what the theory was supposed to explain.

There is a tension here, which is apparent in the suspicion with which certain philosophers (including Stich and Tye) regard what they call linguistic or a priori intuitions.[37] On the one hand, they argue against *psychofunctionalism;* on the other, they rebel against the too-strict requirements of traditional conceptual analysis.[38] Thus Stich suggests

---

37. Tye's goal is to reject the idea that naturalism about the mental requires (even partial) reductionist attempts to find mental essences but nonetheless to defend the view that mentality is a natural phenomenon a) because it is what psychology studies and b) because there is no reason to think that the methods of psychology are different from chemistry, biology, and geology, which support a hierarchy of constitution and realization relationships between higher- and lower-order physical items, and which has as its foundation in the microphysical realm. Thus there is no reason to doubt that "mental states participate in causal interactions which fall under scientific laws, and are either ultimately constituted by or ultimately realized by microphysical phenomena." Tye, "Naturalism and the Mental," 436. But this picture of what licenses naturalism simply presupposes an answer to the problem that the reductionists were trying to solve. How, after all, are any putative realization or constitution relations to be tested if not by appeal to considered judgments about the appropriate use of mental concepts? For an interesting discussion of the importance vs. irrelevance of a priori intuitions regarding a particular philosophical thought experiment, see the forum on "swamp-people" in *Mind & Language* 11, 1 (1996).

38. Tye objects to this doctrine on the grounds that if empirical work is completely unconstrained by a priori intuitions then "just about any answer [as to what constitutes the essence of the mental] will do" (ibid., 427). And as both he and Stich point out, adherents of this view are showing an allegiance to a controversial view about reference.

that an investigation of the body of knowledge that underlies commonsense practices is the business of cognitive anthropologists. Tye, for his part, suggests that whatever the mental is, it is what (empirical) psychologists investigate. In other words, a rejection of classical analysis is for them a rejection of conceptual philosophy.

As I argued above, we can agree with their suspicion if an appeal to armchair intuitions is supposed to involve a mysterious epistemic route to Meanings (Concepts or Propositions) as rules that predate and in some sense determine our linguistic practices. But the same mythical picture, now individualized or internalized, and relying on innate knowledge to solve the epistemological problems of the classical approach, is unfortunately adopted and only slightly modified by the program in philosophy of language that provides the philosophical foundations for the cognitivist approach.

Cognitivism aside, in exploring the two motivations for constructing a theory of content, we come face to face with different accounts of the explanatory power of the mental. Does it lie in hidden causes of some nonmental description that are best uncovered by empirical science? Or is it to be identified in semi-hidden causes of some mental description whose realizations, which subvene on mental states, will one day be uncovered by empirical science? Stich is right that those who are interested in theories of mental content or representation seem not to have made up their minds about this. This is evident in the disagreement about whether "mental representation" is just another term for the propositional attitude states that are alleged to be what we are identifying with these mental expressions—in which case considerations about what would constitute the correct application of these concepts would seem to be in order—or whether mental representations are theoretical concepts that have little to do with the ordinary use we make of expressions containing mental predicates.

And as Wittgenstein and Ryle argue, it is a view that makes it a mystery how our ability to understand one another by attributing mental concepts could have ever come about in the first place if our application of them is so rough.

This brings us to the second reason to construct a theory of mental content: one which would completely jettison a conceptual, philosophical approach. Although those interested in theories of mental content *may* be interested in the role of mental concepts in everyday explanation, they may not. Cultural anthropologists, for example, offer insight into different cultures' concept of disease, say, just as those philosophers who study common-sense mental concepts offer insight into the folk-psychological practices in which they are ascribed. But just as the anthropologist tells us nothing per se about the nature of disease, neither does the philosopher studying folk-psychology tell us anything per se about the nature of mental "properties" that play such an obvious role in our everyday lives. As Stich elaborates:

> On this second account [of what a theory of mental representation is supposed to be], a theory of mental representation doesn't much care about the common-sense conception of mental representation. The intuitions and tacit knowledge of the man or woman on the street are quite irrelevant. The theory seeks to say what mental representation really is, not what folk psychology takes it to be. And to do this it must describe, and perhaps patch up, the notion of mental representation as it is used by the best cognitive science we have available. So on this account, a theory of mental representation begins as part of the cognitive psychology of cognitive science, though it may end up contributing to the conceptual foundations of the science it sets out to describe.[39]

On this understanding, the project explicitly shuns ordinary "intuitions" and instead attempts to say what various approaches in the cognitive sciences take mental representations to be. Is it really to be hoped that this investigation can proceed in a way that is not constrained by considered judgments about the correct uses of expressions containing mental concepts?

39. Stich, "What Is a Theory of Mental Representation?," 355.

The problem with the view that mental representations are theoretical posits that are not expected to meet any constraints at all on common-sense practice is that it leaves us in doubt about what the theory in which these theoretical posits figure is supposed to be about. Thus the proposal that a theory of mental representation begins as part of the cognitive psychology of cognitive science *and is thus* not answerable to common sense, considered judgments about the application of intentional concepts is nonsense.

For according to this use of "mental representation," a theory of mental representation is supposed to help philosophers explain what a cognitive scientist does when she tries to explain . . . what? What is the cognitive scientist supposed to be explaining? Our cognitive capacities, of course, but how do *these* relate to what we *count* as reasoning, deliberation, rationality, knowledge, thinking, calculating, perceiving, and so on? Even if the project is prepared to be revisionist, there will always be an open question—about a particular proposal to revise our use of mental concepts in the light of these theories—whether those concepts would continue to serve our needs if the proposed revisions were to be accepted and guide our use. This ought to provide a fruitful philosophical "research program" that should operate in tandem on the deliverances of any empirical-psychological theory.[40]

If, however, the relation is severed between how we use these concepts and what the empirical/cognitive scientist is supposed to study, then such a scientist will be studying phenomena (and calling them "cognitive capacities" and positing entities called "mental representations") that come about only as a result of her *élucubrations*. If this is indeed a fair way to describe her project, in no sense, then, would she be able to shed much light, let alone be the final arbiter, on the nature of the mind.

40. See Clark, *Microcognition,* for a similar appeal for cooperation, except that I (unlike Clark) am here expressing reservations about one empirical program in particular: the very idea that cognition is a form of information processing.

CHAPTER TWELVE

Ryle's Regress and the Philosophy of Cognitive Science

> In *Final Fantasy* [the first film to have an entirely computer-generated female lead] not only were the mouths of the real actors filmed and digitised, so that the virtual characters' lips could be synchronised with the sound, but every bit of walking, clambering and jumping was first done in a studio by stuntmen, so that the movements could be pasted wholesale on to digital skeletons. . . . The directors have made people do all this, and then thrown the people away. As such some viewers might be forgiven for considering *Final Fantasy* to be a giant con trick . . . the cast of *Final Fantasy* are merely the pixellated ghosts of human movement. It is a melancholy species of shadow puppet theatre.
>
> Steven Poole's review of the film *Final Fantasy: The Spirits Within,* in *New Statesman,* August 6, 2001: 29

1. Ryle's regress objection to the "Intellectualist Legend"—that intelligent activity requires prior theoretical operations—was recognized by Fodor to present a powerful conceptual obstacle to the premise that underlies cognitivist approaches in the sciences. He attempts to thwart Ryle's argument in *The Language of Thought* by accusing him of confusing causal and conceptual explanations and claiming that, by analogy with computers, we can see how the appeal to explicit rules is halted at the first level since second-order rules are reducible to built-in causal processes.

In this chapter I attempt to show that Fodor's arguments against Ryle fail. In the next section, I suggest that Fodor's appeal to the "empirical necessity" of theoretical operations misfires because he is the one who has misunderstood the difference between causal and conceptual questions. In section 3, I argue that the fact that second-order rules are reducible to causal processes shows, not that the regress is halted, but that we cannot consider intelligent activity by analogy with computers. In section 4, I discuss the philosophical motivation for introducing rules into an account of intelligent activity in the first place.

2. In *The Concept of Mind,* Ryle spells out a vicious regress that confronts the Rationalist or "Intellectualist": performing some activity intelligently, rationally, or with reason cannot require prior theoretical operations such as deliberating, calculating, or following rules since these are activities that are themselves performed intelligently, rationally, or with reason. So it must be possible to act intelligently, rationally, or with reason without prior deliberation, calculation, or rule-following. Otherwise one would need to suppose the existence of prior theoretical operations ad infinitum.

In the introduction to *The Language of Thought* Fodor remarks that it is difficult to think of an area of cognitive psychology in which the array of arguments in *The Concept of Mind* would not apply or in which Ryle does not apply them:

> Indeed, it is perhaps Ryle's *central* point that "Cartesian" (i.e., mentalistic) psychological theories treat what is really a *logical* relation between aspects of a single event as though it were a causal relation between pairs of distinct events. It is this tendency to give mechanistic answers to conceptual questions which, according to Ryle, leads the mentalist to orgies of regrettable hypostasis: i.e., to attempting to explain behavior by reference to underlying psychological mechanisms.[1]

---

1. Jerry Fodor, *The Language of Thought* (New York: Crowell, 1975), 5. Fodor's update on the language of thought hypothesis takes most of his earlier arguments for granted. Although he revisits the "pragmatist" criticism of the appeal to explicit rules, his arguments tend to reiterate, but do not much expand upon, the ones he addresses in more

Fodor goes on to say:

> If this *is* a mistake, I'm in trouble. For it will be the pervasive assumption of my discussion that such explanations, however often they may prove to be empirically unsound are, in principle, methodologically impeccable.[2]

Having said this, Fodor spends a few pages addressing Ryle's arguments before developing his own particular version of the representational theory of mind: one that construes cognition as rule-governed computations over syntactically structured symbols or representations.[3] I will be considering these arguments as we go along.

Language-learning, perception, and rational choice are paradigmatic of abilities, according to Fodor, that admit of a cognitive/computational explanation. Fodor's argument for this begins with his presenting the following model as "an overwhelmingly plausible" account of how at least some behavior is decided upon.

> [a.] The agent finds himself in a certain situation *(S)*.
> [b.] The agent believes that a certain set of behavioral options *(B1, B2, . . . Bn)* is available to him in *S;* i.e., given *S, B1* through *Bn* are the things the agent believes that he can do.
> [c.] The probable consequences of performing each of *B1* through *Bn* are predicted; i.e., the agent computes a set of hypotheticals of roughly the form if *B1* is performed in *S,* then, with a certain probability, *Ci.* Which such hypotheticals are computed and which probabilities are assigned will, of course, depend on what the organism (sic) knows or believes about

detail in the earlier work I am considering here. See Jerry Fodor, *Lot 2—The Language of Thought Revisited* (Oxford: Oxford University Press, 2008).

2. Ibid., 5.

3. This discussion of Ryle follows a more extensive treatment in Jerry Fodor, *Psychological Explanation: An Introduction to the Philosophy of Psychology* (New York: Random House, 1968).

situations like *S*.[4] (It will also depend upon other variables which are, from the point of view of the present model, merely noisy: time pressure, the amount of computation space available to the organism, etc.)

[d.] A preference order is assigned to the consequences.

[e.] The organism's choice of behavior is determined as a function of the preferences and the probability assigned.[5]

In a footnote, Fodor admits that the model does not provide a "logically necessary" condition for rational behavior:

> . . . the conceptual story about what makes behavior rational presumably requires a certain kind of correspondence between behavior and belief but doesn't care about the character of the processes whereby that correspondence is effected.[6]

A few sentences earlier in the text, however, conceding that the model is highly idealized, he argues that the most this concession shows is that "the behaviors we produce aren't always in rational correspondence with the beliefs we hold." Here he does seem to suggest that the deliberation model he describes provides a necessary condition for bringing about a rational correspondence between beliefs and action—at least for human beings: when the model is not adhered to, the behavior is not rational. For he goes on to suggest that though angels may be rational by reflex, the model, "or something like it," may be "empirically necessary" for bringing about a rational correspondence between the beliefs and the behaviors of human and other "sublunary" creatures:

> . . . some agents are rational to some extent some of the time, and . . . when they are, and to the extent that they are, processes [like

4. Note that agents are the possessors of the abilities to begin with and then, without argument or remark, the abilities are attributed to organisms.

5. Fodor, *The Language of Thought,* 28.

6. Ibid., 29.

> these] mediate the relation between what the agent believes and what he does.[7]

What started out as an intuitively plausible model of how at least some behavior is decided upon has turned into a theory about what is necessary for rational action in human beings. What occurs in some passages as an empirical hypothesis about underlying mechanisms occurs in others as a covert conceptual claim: it suggests what we correctly count as rational action in human beings. Fodor can no longer appeal to intuitions for this claim, whatever its status, so where is the argument? If the overt conceptual story, as he concedes, merely requires a correspondence between attitudes and behavior and is silent about any processes involved, then angels may indeed be rational without deliberating, but so may human beings. Curiously, Fodor also admits (in the same footnote) that the production of behavior that follows the deliberation model is not *sufficient* for the rationality of the behavior, since the beliefs involved may be superstitious or the preferences perverse, or the computation grossly unsound.

Here we hit upon the matter that bothered Ryle in his arguments against the Intellectualist (and, indeed, Wittgenstein in his discussion of rules). When we consider the (relatively rare) circumstances in which we *do* deliberate as the model suggests, a multitude of things can go wrong. In predicting the probable consequences of a range of behaviors, assigning a preference order to them, determining his choice as a function of the preferences and the probability assigned, the agent is, we assume, acting rationally. Since the ability to deliberate about one's options is itself a higher-order rational ability, it cannot be required for rationality. It cannot be considered necessary because the alleged explanation (the higher-order ability) presupposes the very thing (rational action) it is supposed to explain. It is not that it is merely insufficient and that something else is needed: it is insufficient because the explanandum is presupposed in the explanans. That is why it cannot

7. Ibid., 29.

be considered necessary either. This is the gist of Ryle's infinite regress objection to the Intellectualist Legend.

Let us look again at Fodor's claim. If it is conceded that an agent need not deliberate as the model suggests in order for his performance to count as rational (which, for argument's sake, let us accept lines up with his beliefs) then what does it mean to say that it is empirically necessary? What could it mean to say that human agents must deliberate as the model suggests when it is conceded that even if they were to, doing so is neither necessary nor sufficient for what we would justifiably count as rational action? In order to clarify this, Fodor refers us back to his earlier treatment of the difference between the kind of conceptual story that typically interests analytic philosophers and the type of causal story that interests psychologists. The suggestion is that Ryle, in his impatience with those who attempt to offer causal answers to conceptual questions, is guilty of confusing conceptual with empirical claims, for the fact that there can be a conceptual answer to the question what makes [an] *x* [an] *F* does not rule out the possibility of a causal story as well.

> In general, suppose that *C* is a conceptually sufficient condition for having the property *P*, and suppose that some individual *a* does, in brute fact, satisfy *C*, so that *"Pa"* is a contingent statement true of *a*. Then: (a) it is normally pertinent to ask for a causal / mechanistic explanation of the fact that *"Pa"* is true; (b) such an explanation will normally constitute a (candidate) answer to the question: "What makes *a* exhibit the property *P?*"; (c) referring to the fact that *a* satisfies *C* will normally *not* constitute a causal/mechanistic explanation of the fact that *a* exhibits the property P, although, (d) reference to the fact that *a* satisfies *C* may constitute a certain (different) kind of answer to "What makes *'Pa'* true?" [. . .]
>
> To put this point as generally as I know how, even if the behaviorists were right in supposing that logically necessary and sufficient conditions for behavior being of a certain kind can be given (just) in terms of stimulus and response variables, that fact would not in the least prejudice the mentalist's claim that the *causation* of

behavior is determined by, and explicable in terms of, the organism's internal states.[8]

Of course it is true that conceptual claims involving causal concepts may invite the search for underlying mechanisms. If the concept of a heart is the concept of that which pumps blood throughout the organism, a search for what functions as the pump in different creatures would be in order; the concept of poison invites an investigation as to whether a particular chemical, for example, causes illness or death; and the concept of disease allows, and then may change its contours to accommodate the results of, a search for viruses or bacteria responsible for its symptoms.

Suppose Fodor is right that the concept of rationality requires agents' beliefs and actions to line up in rational correspondence. An explanation of this could be as follows: we ascribe beliefs, etc. as part of the enterprise of making sense of an agent, and we do this by attributing to him beliefs and other attitudes that line up with his actions, as described. Such is the goal of the enterprise; that is why actions and beliefs (etc.) tend to line up.[9] So far there is nothing to suggest that a causal story explaining the rational lineup is to be found inside the individual or his brain.

In order to underline the point, consider Fodor's own discussion of the conceptual and causal explanations that may both be given for General Mills's claim that "Wheaties is the breakfast of champions." The conceptual explanation is that a nonnegligible number of champions eat the wheat-flake cereal; a causal story perfectly consistent with this, Fodor suggests, may advert to the vitamins and special springiness of the flake's molecules.

The problem with this analogy is that the underlying causal story is only consistent with the conceptual one if the "because" in "because a

8. Ibid., 8.

9. Compare: the explanation why so many games of chess end up in checkmate is that such is the goal of the game.

non-negligible number of champions eat it" is a causal one. One of the main aims of Ryle's work is to suggest that mental expressions, among a large number of others, discharge their explanatory role other than by attributing a property (a fortiori a property with causal powers) to an object. Thus, as with the cases for which the Wheaties example is intended to be a model, here too we have been given no reason for looking inside the composition of the cereal in our search for a mechanism that underlies the conceptual claim. For a nonnegligible number of champions may eat Wheaties because they have been paid a lot of money to endorse the product; or because the champions they most admire do; or because they are given the breakfast free during sporting events; or because they like them, and so forth and so on. In these cases, even if the molecules in Wheaties flakes *are* especially springy, this will be completely irrelevant to an explanation of why a large number of champions eat them.

It may be instructive to take a quick look at other powers—"thick" powers we might say—that may be found on a high rung of what Ryle calls "the ladder of sophistication" in order to see how silly it would be to search for so-called "natural" properties or relations to explain them. A penny has a certain purchasing power; a football has score-enabling power; and a bishop in chess has the power to move diagonally across the board. If an anthropologist from Mars were to wonder what explains these respective powers he would be making a mistake to take the penny, football, or bishop back to his laboratory to search for their physical realizations. This is obviously not to suggest that the respective powers are supernatural: he is just looking in the wrong place for the explanation he seeks. Nor is this to deny that the penny, the football, or the bishop have interesting physical properties, perhaps even properties that enable them to perform these roles. If a penny were not a certain size and weight it could not be used in transactions for gumballs; if the football were not shaped exactly so, a Beckham could not bend it; and if a chess piece were made of ice it would not survive competition on a very hot day. Nor is this to deny that the penny is ripe for a mineralogical examination, a historical-numismatic examination, an investigation for fingerprints, for counterfeiting, and so on indefinitely. But none of these investigations are going to shed much light on its purchasing power. A credit card would be a more suitable candidate, for its magnetic strip and chip encode information that plays an even more complex role in the sort of economic transactions in which it trades. But however complex the credit card, we will not find an answer to how it gets its purchasing power without adverting to the banking institutions and economic environment in which it plays its role. The human brain is, by all accounts, the most complex and wonderful object in the world. But so far we are left with no reason to suppose that an answer to the question about why our beliefs line up with our actions is one that can be given by looking at second-order properties that supervene on matter that is to be found inside the agent's skull.

Like the alien anthropologist who takes the penny home to examine it, the cognitive psychologist may well be looking in the wrong place for an explanation of the agent's rational powers. To insist that the purposive elements of the mental phenomena under examination—and thus the normative dimension along which they can be assessed—may be partly explained by special content-bearing states supervenient upon the brain is as misguided as insisting that the normative dimension upon which a game of football can be assessed may be partly explained by special score-enabling properties (which, in a "long-arm" version, reach out into the world) that somehow emerge from the molecules of the football.

I suggest there is no reason to believe that deliberation—"or something like it"—is necessary for rational action in human beings. Although we deliberate as the model suggests some of the time, Ryle's argument shows that the supposition that we must do so leads to a vicious regress. To anticipate what is to come: to show that a hypothesized mechanism which is alleged to play the role of, or go proxy for, deliberation does *not* lead to regress fails to meet the point of Ryle's objection, which is that deliberation of the kind suggested by the model is not, because it cannot be, conceptually required for rational action. It seems that Fodor, and not Ryle, is guilty of confusing conceptual with empirical claims.[10]

---

10. There are various confusions and conflations in Fodor's discussion of Ryle both in *The Language of Thought* and in his earlier *Psychological Explanation*. Part of the difficulty is

3. THE COGNITIVE-PSYCHOLOGICAL models Fodor describes (those of deliberation, perception, and hypothesis-testing for language-learning) presuppose that the agent has access to a representational system of considerable richness. This system, he argues, must share a number of characteristic features of real languages, and it requires an ontological commitment to the processes and states ascribed by that model.

> . . . [D]eciding is a computational process; the act the agent performs is the consequence of computations defined over representations of possible actions. No representations, no computations. No computations, no model.[11]

When he considers our understanding of the language of representations over which the alleged computations are performed Fodor

---

that though Fodor (rightly) identifies Ryle as working in a Wittgensteinian spirit, he also (wrongly) identifies Ryle's program with that of the logical behaviorist: i.e., one, who, according to Fodor, hopes to show that "logically necessary and sufficient conditions for behaviour being of a certain kind can be given in terms of stimulus and response variables." Fodor, *Psychological Explanation,* 8. He also (wrongly) suggests that the appeal to criteria implies a "criteriological" theory of meaning which he construes as a type of (property) cluster theory (ibid, n5).

Sometimes when philosophers invoke the distinction between logical, conceptual, metaphysical, and physical possibility or impossibility, they take the conceptual story to involve what we can consistently imagine to be the case in logically possible worlds that may be at a distance from ours, vis à vis the (or our) laws of nature. On this view, the use of predicates in the relevant domain of discourse tends to drop out of the picture. Conceptual cartography, by contrast, traces the inflections of meaning or elasticities of significance of the relevant concepts: in this case, the term "rational" and its cognates. To a philosopher engaged in such a project, it would not make sense to say that one could explore a concept (such as rationality) and ignore the way the relevant predicates are used. So another mistake Fodor makes in discussing Ryle involves mixing up Ryle's view of philosophy as conceptual cartography, in which concepts are construed as double abstractions from sentences performing their various jobs, with those philosophers (Frege, early Russell, and Moore) who suppose concepts to be, in the Platonic tradition, independently existing ideals (whose "essences" can be unpacked by necessary and sufficient conditions for the concepts' application) at which our natural (imperfect) expressions merely gesture. Ryle's criticism of this approach began with his first articles in the late 1920s and continued for the rest of his career.

11. Fodor, *The Language of Thought,* 31.

addresses the threat of regress. For if understanding a predicate of English requires representing the extension of that predicate in a language that one already understands, then what about the predicates of the metalanguage? If understanding an English predicate requires representing its extension in a language then understanding the predicate of a metalanguage presumably requires representing its extension in some meta-meta-language, ad infinitum.

Here, Fodor allows that what we may count as "understanding the predicate of a meta-language" may *not* require representing its extension: "a sufficient condition [for this] might be just that one's use of the predicate is always in fact conformable to the truth rule."[12] To put this in Rylean terms (in a way Fodor does not) it seems that he allows that understanding a predicate in a metalanguage may be a matter of knowing this language *by wont;* that is, without prior theoretical operations. The obvious rejoinder, which Fodor acknowledges, is that he may as well have allowed that learning English is a skill that is acquired through training and practice: a kind of knowledge-how or knowledge-by-wont that underlies any explicit higher-order practice of following a meaning-rule.

Fodor fails, after a long discussion, to meet this objection. He compares the language of thought (Mentalese) and a natural language such as English with the two languages used by computers: the input/output (programming) language and a machine language which the machine is *built* to use.

> Roughly, the machine language differs from the input/output language in that its formulae correspond directly to computationally relevant physical states and operations of the machine: The physics of the machine thus guarantees that the sequences of states and operations it runs through in the course of its computations respect the semantic constraints on formulae in its internal language. What takes the place of a truth definition for the machine language is simply the engineering principles which guarantee this correspondence.[13]

12. Ibid., 65.
13. Ibid., 66.

Although this kind of correspondence may also hold between physical states of the machine and formulae of the input/output code, this could be effected only by first translating them into the machine language. It is in this sense that the machine "follows" the input/output code but only acts in accordance with its machine language. The idea seems to be that while genuine rule-following occurs in the programming language, mere rule-conformity occurs in the machine language. (I shall be challenging this idea in the next section.)

This thought is developed further in a footnote in which Fodor considers what is involved in thinking of an organism as a computer. It is wrongheaded, he contends, to think of the nervous system as issuing commands which must be "read" and translated into action or behavior by a further system that intervenes between the efferent nerves and the effectors:

> On the contrary, what is required is just that the *causal* properties of such physical events as are interpreted as messages in the internal code must be compatible with the *linguistic* properties that the interpretation assigns to those events. Thus, if events of the physical type *P* are to be interpreted as commands to effector system *E,* then it better be the case that, *ceteris paribus,* occurrences of *P*-events are causally sufficient for activating *E.* (*Ceteris paribus* means: barring mechanical breakdown and barring events interpretable as overriding countercommands to *E.*)[14]

In his next footnote, Fodor responds to the critic who suggests that if one is willing to attribute regularities in the behavior of organisms to rules that they unconsciously follow, one may as well attribute, say, Kepler's laws to planets in pursuit of their orbits. The point of the critic's remark, Fodor says, is to suggest "that the only *real* case of rule following is conscious rule following by articulate organisms."[15] Other organisms merely act in accordance with rules; they do not follow them.

14. Ibid., 74.
15. Ibid., 74.

Fodor replies to this imagined critic that what distinguishes unconscious rule-following from the rule-conformity exhibited by planets is that "a *representation of the rules* [*organisms*] *follow constitutes one of the causal determinants of their behavior.*"[16] Such is not the case for planets: "At no point in a causal account of their turning does one advert to a structure which encodes Kepler's laws and causes them to turn."[17]

This is an ill-considered criticism of the position, which I shall have something to say about in the next section. In any case, Fodor's tactic is clear: if rules can be seen to be encoded or represented in the structure of the system and constitute one of the causal determinants of behavior, the organism (or its subsystems) can be said to be following them. Indeed, this move is now a familiar one in cognitive psychology and computer design, for both are faced with what is known as the homunculus problem, which, it is generally accepted, can be avoided. Indeed, some understand this threat to be an articulation of Ryle's regress argument, so it will be worth a short digression to examine this problem in a bit more detail.[18]

The cognitive scientist, whose job is to explain some particular complex form of cognitive behavior, will attempt to construe the behavior as an information-processing task. The classical cognitive scientist will construe this, following Turing, as involving rule-governed computations over syntactically structured symbols. In order to say how the information-processing task is accomplished in the system in question, the cognitive psychologist uses a method called forward engineering or functional analysis. The aim of the analysis at this mid-level is to uncover the particular algorithm that is assumed to be "used" by the system to complete the information-processing task that was defined by the higher-order computational analysis. In other words, since it is assumed that some effective procedure or other is implemented by the system to tackle the information-processing task identified at the computational level of analysis, the goal of forward engineering or func-

16. Ibid., 74.

17. Ibid., 74.

18. The discussion below in the text follows Michael R. W. Dawson, *Understanding Cognitive Science* (Oxford and Malden, MA: Blackwell, 1998).

tional analysis is to identify that procedure. Whereas the programmer or designer of some device will amass rules, heuristics, and brute procedures for producing the correct responses for some task, a (classical) cognitive scientist will assume that his subject is using some combination of such procedures and attempt to determine which ones, or "re[-]create the program that is producing the behavior."[19] This is the goal of functional analysis. But if the goal is to analyze or divide complex functions into subfunctions, the danger is that they will never end.

> It would appear that functional analysis leads us directly into Ryle's Regress; each function that we propose to explain an earlier black box will rear its head as an ugly homunculus. We wind up with an infinite proliferation of unexplained functional terms.[20]

Forward engineering must be constrained so that each functional decomposition is decomposable into simpler functions. There has to be some principled reasons for supposing that functional decomposition must stop. And finally, it must be clear how the set of functions that exist at the end of forward engineering can be physically implemented. "It is only with this third claim that a classical cognitive scientist can abolish the ghost from the machine."[21]

Although Dennett expresses doubt about various aspects of Fodor's project, he explicitly praises this way of avoiding the homunculus problem in his review of *The Language of Thought:*

> . . . perhaps the *prima facie* absurd notion of self-understanding representations is an idea whose time has come, for what are the "data structures" of computer science if not just that: representations that understand themselves? In a computer, a command to dig goes straight to the shovel, as it were, eliminating the comprehending and obeying middleman. Not *straight* to the shovel, of

19. Ibid., 109.
20. Ibid., 156.
21. Ibid., 157.

> course, for a lot of sophisticated switching machinery is required to get the right command going to the right tools, and for some purposes it is illuminating to treat parts of this switching machinery as analogous to displaced shovellers, subcontractors and contractors. The beauty of it all, and its importance for psychology, is precisely that it promises to solve Hume's problem by giving us a model of vehicles of representation that function without exempt agents for whom they are ploys. Alternatively, one could insist that the very lack of exempt agents in computers to be the users of the putative representations shows that computers do not contain representations—real representations—at all, but unless one views this as a rather modest bit of lexicographical purism, one is in danger of discarding one of the most promising conceptual advances ever to fall into philosophers' hands.[22]

I shall be arguing that the suggestion that computers do not contain real representations is *not* a modest bit of lexicographical purism: indeed, the ontological status of Fodor's hypothesized language of thought is at stake (as is, incidentally, the realism of the Representational Theory of Mind, as well as that of the content-bearing mental states that are alleged to play a causal/explanatory role in functionalist philosophies of mind). Here, let us note that it is not at all clear how the digression on machine code language helps Fodor's argument. Fodor turned to computers when asked why we could not say "we're just built that way" one level earlier—to explain how we learn the English predicate "is a chair"—thus making the retreat from the natural language to the inner language unnecessary. After introducing the difference between programming language and machine code, Fodor returns to this question and replies that he agrees (presumably with Wittgenstein) that explanation has to stop somewhere, "but it doesn't have to—and better not—stop *here*":

22. Daniel Dennett, "A Cure for the Common Code?," (1977), in Daniel Dennett, *Brainstorms—Philosophical Essays on Mind and Psychology* (Montgomery, VT: Bradford Books, 1978), 102.

> The question of how we learn "is a chair" *does* arise precisely because English *is* learned. The question of how [a predicate of the metalanguage] is learned does not arise precisely because, by hypothesis, the language in which [the predicate] is a formula is innate.[23]

This extraordinary response simply begs the question; it assumes but offers no support for the claim that we need to posit a metalanguage to explain how English is learned. The model can only be considered a "hypothesis to the best explanation" if we have some reason to believe that an explanation of the kind proposed is required.

The same criticism can be raised against Fodor's language-learning hypothesis. Presumably Fodor would agree that the conceptual story for (what we correctly count as) learning a (first, natural) language does not require *explicit* hypothesis-testing. If we do on occasion test a hypothesis for the application of a predicate it is either without using language to do so or, if we do use language, a fair bit must have already been learned. After all, the same regress arguments can be marshaled to show that hypothesis-testing is not sufficient, because of the possibility of error. And, of course, the (natural) language in which explicit hypothesis-testing would be couched presupposes the very thing the hypothesis-testing is supposed to explain. So it cannot be necessary either. Thus there is no motivation for supposing the existence of an underlying mechanism that subserves the role of explicit hypothesis-testing, for it has just been conceded that hypothesis-testing is neither necessary nor sufficient. Thus the suggestion that there is a nonexplicit, innate, error-free, hypothesis-testing mechanism that does not threaten regress is completely unmotivated. In other words, to argue as Dennett does that the regress is not vicious (because the homunculi become "stupider" and "stupider" until they finally drop out of the picture) would at most block the argument from absurdity and allow that prior—mechanical—operations are not ruled out. But this would still yield no support at all for the covert conceptual claim that such

23. Fodor, *The Language of Thought*, 67.

mechanical operations are necessary to explain the ability of human beings to act rationally or understand language.

Fodor's response to the objection that the retreat from natural language to Mentalese is not necessary simply begs the question against those who think that learning English may involve training, practice, and the (innate, if you wish) capacity to catch on to similarities that are natural in the circumstances.[24] Nothing in the fact that English is learned requires that this learning must be effected via mechanized hypothesis-formation which in turn would require a metalanguage.

I have argued that the point of Ryle's regress argument against the Rationalist or Intellectualist was to show that prior theoretical operations cannot be considered necessary for rational action, language learning, or intelligent abilities in general, for the supposition leads to absurdity. The conclusion is that it must be possible to act rationally, understand a language, or act intelligently without participating in what are in effect higher-order activities requiring the same abilities.

Fodor seems to agree that prior theoretical operations are not conceptually required, but maintains they are "empirically necessary" for human beings. This is confused. If such higher-order activities were conceptually required, it would make sense to look for mechanisms that subserve these operations in human beings. Since they are not, it does not. Fodor thus seems to be making the covert conceptual claim that such operations are necessary: not for angels, but for human beings and other "sublunary" creatures. We have been given no reason,

---

24. Rey also gives short shrift to the objection by parodying the attitude—crystallized in Wittgenstein's comment that explanations have to come to an end somewhere—by saying :

> [O]ne could . . . [rationally] . . . never ask for explanations of anything at all: we *could* just say that it is a natural capacity of lightning to burn what it strikes. The question is whether we *can also* rationally ask for slightly deeper explanations, and, if we can, what those explanations might be. (Georges Rey, *Contemporary Philosophy of Mind—A Contentiously Classical Approach* [Cambridge, MA, and Oxford: Blackwell, 1997], 5)

This also is too quick, since Wittgenstein's discussion of rules, like Ryle's battle with the Intellectualist, constitutes a sustained argument for the limits of rule-following explanations or explanations requiring prior theoretical operations.

however, for believing this claim, other than that it would make theoretical sense of work in the cognitive sciences. On the contrary, if the operations are, as Ryle envisaged, higher-order rational activities, then they are ones that assume the ability to be explained, and we have been given an overwhelming reason for rejecting the claim.

In assimilating Ryle's regress with the homunculus problem in cognitive science and computer design, Fodor attempts to maintain that the regress is not a threat because it is not vicious. I have argued in response that even if this assimilation were in order, it could at most block the argument from absurdity by retreating to some other kind of operation—one that is not a higher-order intellectual activity. Yet we have no argument that such an operation is necessary. Indeed, with this move to a different kind of (because now mechanized) operation, we are bereft of the original "overwhelmingly plausible" account of how some behavior is decided upon or, generalizing to cover the case of hypothesis-testing, how some languages are learned, since *that* model has been replaced by one with very little in common with it, as we shall see. The result is that we are left with no positive reason to believe that mechanized "deliberation" or "hypothesis-testing" is necessary for rational action and language-learning in any creature.

To be sure, some stretch of activity may be photographically or telephonically indistinguishable from that which is assessable as intelligent, rational, meaningful, etc. We still need a way of distinguishing these cases and for this we need to appeal to norms, or the criteria in virtue of which such activities may be assessed.[25] Ryle is right that acting intelligently, etc. does not require that the agent casts a sideways glance at rules that embody these criteria. I have argued above that

---

25. The intentions and propositional attitudes of the agent are indeed relevant in distinguishing these situations. But to attribute an intention is not a matter of naming a particular kind of mental state, roughly in the brain, that takes propositional content as its object. It is to deploy a linguistic tool—several steps up the "ladder of sophistication" of mental discourse—that puts a marker down on just those assessment conditions that are relevant. See "Reason-Explanation and the Contents of the Mind," "Reasons as Non-Causal, Context-Placing Explanations," and "Pain, Polio, and Pride: Some Reflections on 'Becausal' Explanations" (chapters 6–8) for more on the explanatory function and logic of intention and other propositional attitude attributions.

the supposition that these rules are hardwired is unmotivated; below I shall argue that mechanizing the rules renders the norms which they embody unusable as a tool for assessment.

4. Why would anyone think that prior deliberation—or something like it—is necessary for rational action? What motivates the rationalist, for example, to think that intelligence in general requires prior theoretical operations?

The Rationalist, or Intellectualist, might argue as follows: what distinguishes one bodily movement from another imperceptibly different movement (in one sense of "imperceptible") is the fact that one is intentional, the result of agency, performed for reasons: the other not. On the rationalist construal, this difference amounts to the occurrence of a nonperceptual, mental feature (intention, deliberation, reason) that plays a causal role issuing in behavior. Similarly, what distinguishes the meaningfulness of a person's utterance from a phonetically similar sound made by, say, a parrot is the addition, in the first case, of a mental act of meaning. What distinguishes acts of hearing from acts of listening is the mental accompaniment of understanding. What distinguishes an inference from a mere string of statements is that the first was, but the second was not, made "under the influence" of the rules of logic.

According to Ryle, this construal of what is required for rational action, language understanding, or intelligence may have derived from Plato's doctrine of the tripartite soul in which Intelligence is a special faculty by which internal acts of thinking—particularly the consideration of regulative propositions—are exercised. This construal of what is required seemed to be encouraged by the Enlightenment idea that mathematics and natural science set the standard as human accomplishments. Impressed by the analogy, the Rationalist supposed that it was the capacity for theorizing that constituted the intellectual excellence of man, together with

> the idea that the capacity to attain knowledge of truths was the defining property of a mind. Other human powers could be classed as mental only if they could be shown to be somehow piloted by the intellectual grasp of true propositions. To be rational was to be able

> to recognize truths and the connexions between them. To act rationally was, therefore, to have one's non-theoretical propensities controlled by one's apprehension of truths about the conduct of life.[26]

As we have seen, however, since theorizing is itself an intellectual ability it cannot be required for intelligence, on pain of regress. What of the suggestion that grasp of truth or of regulative propositions controls one's nontheoretical propensities in a merely mechanical way? This would at least seem to eliminate the threat of regress. The trouble with this suggestion is that two different kinds of explanation are conflated: the first, an explanation by appeal to standards or norms that are codified in the performance-rules that govern some activity or practice; the second, an explanation by appeal to causal relations or to the laws of nature which are supposed to subsume them.

Rules have different explanatory functions from those of natural laws. The rules that govern rational action or logical inference, for example, help us to identify what is incorrect in certain performances and correct in others. Such explanations function by showing how the performance attains or fails to attain the standards or norms set for the practice. These are explanations relatively high up Ryle's "sophistication ladder." They are descriptions of the referee, theorist, or tactician.

If someone (as I once did, when very young) puts a cup of milk into a mixture for oatmeal cookies, and wonders why the consistency is wrong, pointing to the recipe that says "add 1T milk" is one way of diagnosing the problem. If someone who cooks by intuition tosses into a mixture of sugar, butter, and oatmeal a small amount of milk, and the cookies are good, then measuring the amount and writing it down as part of a recipe could be part of a useful second-order practice for those who want to cook but do not have the natural talent. Similarly, if someone is confused by a complicated argument, showing that the argument, once symbolized, is not valid would be one way of showing what is wrong with it. Indeed, explaining that it conforms to an

---

26. Gilbert Ryle, *The Concept of Mind* (1949) (London and New York: Routledge, 2009), 15.

accepted rule of inference is a way of showing that it is valid. But, as Ryle says, the rules of logic, like the rules of etiquette or the rules of cooking are performance rules: "only performances can be or fail to be in accordance with them."[27]

This, incidentally, is why the appeal to Kepler's laws mentioned earlier was inapt. The point of the critic's remark, recall, was to suggest that the only *real* case of rule-following is conscious rule-following by articulate organisms: other organisms merely act in accordance with rules; they do not follow them. Fodor replied that what distinguishes unconscious rule-following from the rule-conformity exhibited by planets is that "a *representation of the rules* [*organisms*] *follow constitutes one of the causal determinants of their behavior.*" Such is not the case for planets: "At no point in a causal account of their turning does one advert to a structure which encodes Kepler's laws and causes them to turn."[28] Fodor, however, is conflating laws and performance rules. Kepler's laws are not performance rules; it makes no sense to talk about them being followed even by a person who is perfectly capable of representing them, let alone a planet that is not.

Performance rules, do not, as Ryle reminds us, debar mistakes from happening. In this way they are different from laws of nature or those events which are supposed to be causally determined, which

> rule out certain imaginable conjunctions of happenings or states of affairs in quite a different way. Hence while there can and do occur breaches of logical rules, there cannot and do not occur breaches of laws of nature. It makes sense to speak of someone obeying or disobeying a performance-rule, none to speak of things disobeying or obeying laws of nature.[29]

The rules that govern normative practices are performance rules. I suggest that these rules be construed as encoding norms which allow

27. Gilbert Ryle "Why Are the Calculuses of Logic and Arithmetic Applicable to Reality?," (1946), in *Collected Papers,* vol. 2 (London: Routledge, 2009), 241.
28. Fodor, *The Language of Thought,* 74.
29. Ryle "Why Are the Calculuses. . . ," 239.

an observer to pinpoint whether a particular performance has lived up to a standard, *whether or not the one credited with the ability to participate in the practice casts a sideways glance at these rules.* This, I claim, is the upshot of the regress argument: it is possible to credit someone with an ability to participate in a practice even if he is not guided by performance rules. It has to be so: consulting performance rules does not guarantee success or that the performance will live up to the standards which the rules attempt to codify. The second-order practice guarantees nothing because each of the constituent, subordinate actions involved in following a rule, let alone in representing it, can go awry. The rules represented in a logic book, for example, may have been misprinted or mistaken; they may be misread, misinterpreted as applying to a certain situation and not another; once consulted they may be ignored, or may not anticipate novel circumstances, and so forth and so on.

Fodor's imagined critic was half right to say that it only makes sense to attribute the ability to follow a performance rule to conscious, articulate organisms, but the "organism" had better be a full-fledged rational agent since following a performance rule (as opposed merely to acting in accordance with it) involves its own series of rational actions. Indeed, though we may describe someone as following a performance rule if he tells or shows us that he is, telling and showing are yet other (third-order) actions higher up on the sophistication ladder from following the rule, so it cannot be required for it.

Fodor's argument began by giving a model of explicit deliberation that he suggested was a plausible account of how at least some behavior is decided upon. But this model, which involves a host of constituent, subordinate actions exploits principles, which, like many of the performance rules with which we are familiar, have come about in the first place because theorists crystallize them from extant practices, most of which are constantly evolving. Just as a recipe writer will make decisions (with a particular audience in mind) about what moves in a chef's performance are important for the production of a dish so will a mapmaker take a look at what the villagers do automatically in order to provide information of a different kind to one who does not know his way about.

On this way of looking at things, the practice comes first; the codification of the practice comes later. As Ryle reminds us:

> There were reasoners before Aristotle and strategists before Clausewitz. The application of rules of reasoning and strategy did not have to await the work of their codifiers. Aristotle and Clausewitz were, in fact, only able to extract these rules, because they were already being applied. The crystallisation of performance-rules in rule-formulae is, in some cases, not the condition of their being applied [i.e., not necessary in order for something to be an observance of them] but a product of studies in the methodology of the practices in which they have already been applied.[30]

This is why Ryle suggests that when performance rules are applied (as opposed to followed) this is because of the intelligence of the theorists—that is, it is due to the good judgment of the logician, recipe-writer, or mapmaker who has distilled the rules (with his particular audience in mind, as well as his particular purposes for distilling them) from those performances that are deemed successful.[31]

How would we decide if someone is following a performance rule as opposed merely to acting in accord with it?[32] Among the features we would point to in justifying our description of someone as perform-

---

30. Ibid., 243.

31. For more discussion see "Real Rules" (chapter 4) and my "Rethinking Ryle: A Critical Introduction to *The Concept of Mind*" in Gilbert Ryle, *The Concept of Mind* (London: Routledge, 2009), ix–lvii.

32. In this discussion I am, in effect, suggesting that "following a rule" should be reserved for actions that involve the second-order activity of consulting a performance rule. On this way of speaking, the fact that one acts in conformity with the norms, or acts in accordance with a performance rule, will sometimes suffice for deeming his action to be intelligent, etc., and sometimes not, for example, when the action comes about by accident. I have not discussed the relevant considerations here. Others (notably Wittgenstein and Ryle) seem to allow two senses of "following a rule" or "applying criteria" so that one may be following a rule or applying criteria in the sense that they are performing intelligently, etc., without following a rule or applying criteria in the sense of casting a sideways glance at a representation of that rule or codification of those criteria.

ing the second-order task of following a performance rule would be the existence of some representation or expression of that rule. Other features we would expect to find would be the subordinate actions in which following a rule consists: for example, consultation of the rule-representation, and the application of it to the appropriate situation in action. Yet, if a rule can be consulted it can be misread or misunderstood; if it can be applied it can be misapplied. The existence of these constituent, subordinate actions, and the possibility of error that they entail, disappear from the mechanized model; and yet genuine rule-following, and not mere rule-conformity, is supposed to be essential to the model.

This discussion of performance rules gets, I think, to the heart of "the normativity objection" against "naturalistic" theories. The kind of explanation in which we trade when we advert to norms that govern a practice is inextricably tied up with the idea that performances within that practice may fall short of the norm as well as conform to it. The normativity of these practices and the possibility of error are two sides of the same coin.

Indeed, if the goal of computer design is to build a machine that conforms to the norms that govern some practice, why would we need to build it to *consult* performance rules—ones in accordance with which it is guaranteed (barring mechanical failure) to act? Why not merely build it to conform to the norms encoded in these rules? And why should Nature design our brains to *consult* performance rules—ones in accordance with which she guarantees, barring mechanical failure, we will act? Why not merely build our brains to cotton on and eventually to conform to whatever will turn out to be the relevant norms?

The computationalist is committed not only to representations ("tokens" of beliefs, etc.) over which computations are performed; she is also committed to the representation of the rules that govern these computations. If we construe the organism as a Universal Turing Machine, then the mental representations over which computations are performed correspond to the "data" symbols on the tape; the representation of the rules that govern these computations correspond to the machine table of the particular Turing Machine that the Universal Turing Machine is mimicking—or, in short, the particular program computing the particular data. In order to block the suggestion that if the rules themselves are represented there must be further rules explaining how to interpret them, the suggestion is that the implementation of the rules that govern the computations are built into the machine code (they correspond to the Universal Turing Machine's machine table which is built into the head or scanner). Computationalists are obliged to mechanize the rules governing the higher-order ability (i.e., they are obliged to maintain that the rules that govern deliberation and hypothesis-testing are built in) to avoid the explanatory regress. But they insist that rule-representations are being genuinely followed, and not merely instantiated: this is required by the thesis that human beings are Turing Machines and are not mere instantiations of them. But once the rules are "mechanized"—or viewed as causal determinates of the behavior they are alleged to explain—their role as codifications of norms or as the tools used to guide the first-order performance evaporates.[33] This, indeed, is explicit in Fodor when he claims that all that is required for rule-following is just that "the *causal* properties of such physical events as are interpreted as messages in the internal code must be compatible with the *linguistic* properties that the interpretation assigns to those events."

Indeed, the computer analogy lets the representationalist down at just this point. I asked rhetorically why we would need to build a computer that consults performance rules and is guaranteed (barring mechanical failure) to act as they require? Why not merely build it to conform to these rules? But this is of course what we do. Computers do not *compute* rule-representations if this means that they consult performance rules as to how the computations should be effected. Rather the computers' mechanism is built so that certain sequences of operations are applications say, of an inference rule, which, to paraphrase Ryle, is a fact about the efficiency and intelligence of the programmer or designer and not the computer. Nor do computers manipulate *representations* if

33. As Wittgenstein objects when his interlocutor suggests that a meaning rule (or ostensive definition) could be set up for an (allegedly) private object, "What is this ceremony for? For that is all it seems to be!" Ludwig Wittgenstein, *Philosophical Investigations*, G. E. M. Anscombe, trans. (Oxford: Blackwell, 1953), §258.

these are to be construed as symbols with content or meaning; for the whole point of syntactic operations is that their semantic properties are inessential.

When Dennett suggests we ignore what he calls a "rather modest bit of lexicographical purism" he is asking us to disregard an illegitimate equivocation on the expression "representation," which makes all the ontological difference. If the symbols are performance rule-formulae and function as such, then the use is legitimate. But for them to function as such, not only must they be interpret*able* as rule-formulae, they must be so interpret*ed,* and this interpretation must guide the interpreter's performance.

In other words, and to repeat, we may agree to call uninterpreted, but interpretable, symbols "representations" as Fodor and Dennett suggest, but if we agree to stretch the meaning of "representation" in this way, and we extend the computer analogy to the case of human agents, it would present the computationalist with a renewed homunculus problem. For if physical events are interpreted as messages in the internal code, which are in turn compatible with linguistic properties that "*the interpretation* assigns to them," an interpreter, it would seem, is still required to do the interpreting. If however, the rules that govern computations are only represented in the weak sense that the causal processes may but need not be so interpreted, then this is to admit that the rule-representation is not necessary and that the uninterpreted causal processes are sufficient for explanation. This collapses the rule-following/rule-conformity distinction and threatens the ontological status of the rule-representations alleged to govern the computations that operate over them.[34]

The Intellectualist had required the existence of higher-order practices in order to explain the intelligence of lower-order ones; this pre-

34. Although I cannot discuss this here, a similar argument can be run for the first-order representations ("tokens" of beliefs over which the computations are allegedly performed). For the same vehicle-cargo model (as Ryle calls it) is adopted by any view that conceives mental concepts to signal the existence of states or events (the "attitudes") which carry propositions or meaning as content. This means that the representationalist as well as the functionalist is threatened.

sented him with a vicious explanatory regress. The computationalist, in an attempt to avoid the regress, collapses these practices into an algorithm in which there is no genuine latitude as to how the "rules" of the programming language are to be implemented.[35] Whereas the Intellectualist is guilty of requiring the existence of second-order practices which pushes back the explanatory problem, the computationalist is guilty of forgetting that the performance rules belong to the second-order practice, and if they figure at all, they figure for the computer designer. Dennett admits that sophisticated switching machinery is required to get the right command going to the right tools, and that it may be helpful to treat parts of this switching machinery as analogous to displaced shovelers, subcontractors, and contractors. But the importance of this remark is unnoticed. If it is helpful to treat parts of the switching machinery thus, it is for the benefit of the designer. For if any rule-following occurs here, as opposed to mere rule-instantiation, it is the designer who executes it. She is the one deciding how the rule is to be interpreted and implemented; it is she who will have to anticipate new applications.

Could it be that because there has been some success in building expert systems in limited domains, some philosophers have been over-excited by the possibility that we ourselves are organic examples of similar computational systems? The fact that expert systems can be built gives us no reason to believe, however, that we are information processors of a similar sort. The designer of the system is distilling performance rules and heuristics from some practice in which norms or standards that help identify correct or successful moves can be discerned.[36] In building a machine that instantiates the program, she is in

---

35. Even the representationalist is forced to say that mental representations have their content intrinsically in order to avoid a homunculus threat. In functionalism, the crux of the problem comes when we wonder whether it is the mental (semantic) or merely causal properties that play an explanatory role.

36. This is clear to see in PDP architectures, considered by some to be competitors to other psychological models as a means of explaining psychological data. In connectionist systems learning rules are used to train networks to give the "right" outputs, given the inputs, by adjusting connection weights. According to Dawson, in order to make the transition from computer design to cognitive science, the connectionist must deduce the

effect making decisions about how the rules are to be followed. Just like the directors of *Final Fantasy,* these designers have looked at what people do to build their expert systems. The problem is that philosophers who are thus inspired have attempted to throw away the community of rule-conformers and rule-followers by supposing that all of this may be contained within the individual's brain as a part of Nature's design. But computer designers will be happy to admit the artificial machines they build are pixelated ghosts of human practices.

Far from being "natural," the rule-following systems that cognitivist-philosophers imagine us to be are, I fear, just the same.

algorithm that the trained network is "using" and give some evidence that this is biologically sound. What counts as "right" during training, however, is decided by the trainer. If PDP architectures are to be models for human cognition, do connectionist cognitive scientists suppose, with their classical colleagues, that what counts as "right" for humans is determined by the "natural" environment, independently of human practices?

PART FOUR

Self-Knowledge

CHAPTER THIRTEEN

Some Constructivist Thoughts about Self-Knowledge

1. HOW ARE WE to account for the authority granted to first-person reports of mental states? What accounts for the immediacy of these self-ascriptions—the fact that they can be ascribed without appeal to evidence and without the need for justification? A traditional, Cartesian conception of the mind, which says that our thoughts are presented to us directly, completely, and without distortion upon mere internal inspection, would account for these facts, but there is good reason to doubt the cogency of the Cartesian view. Wittgenstein, in his later writings, offered some of the most potent considerations against the traditional view, and contemporary philosophy of mind is practically unanimous in rejecting some of the metaphysical aspects of Cartesianism. But anyone who repudiates Cartesianism shoulders the burden of finding another way to accommodate its apparent epistemological strengths.

Crispin Wright has suggested that Wittgenstein's rule-following passages are specifically concerned to question the idea that our ability to avow our thoughts is epistemically grounded. Wright sees in Wittgenstein an argument for a nondescriptive view about our relation to our own mental states, which, if plausible, would save the phenomenological immediacy of self-ascription, and the practice of granting authority. I would like to consider the view that Wright recommends. I shall argue that it is a merit of Wright's view to allow for an element of creativity in self-ascription, but not at the cost of jettisoning the standards to which we are held accountable when we self-ascribe, and that a plausible account of our ascriptive practices must accommodate both of these features.

The fact that we often self-ascribe directly and without appeal to evidence is recoverable on a view that takes thought content to be self-ascribable as part of an imaginative or creative skill whose standards can be extracted from looking at what we do when the attribution requires reflection or justification. In order to get a clear view of the constraints that are imposed on self-ascriptions, let us consider a situation where an appeal to "evidence" would be indicated. Consider the following passage from Jane Austen's *Emma,* which describes Emma's reaction to Harriet's declaration of love for Mr Knightley.

> Emma's eyes were instantly withdrawn; and she sat silently meditating in a fixed attitude, for a few minutes. A few minutes were sufficient for making her acquainted with her own heart. A mind like hers, once opening to suspicion, made rapid progress. She touched—she admitted—she acknowledged the whole truth. Why was it so much the worse that Harriet should be in love with Mr Knightley than with Mr Churchill? Why was the evil so dreadfully increased by Harriet's having some hope of return? It darted through her, with the speed of an arrow, that Mr Knightley must marry no one but herself![1]

Recall the background: Emma had taken the socially and financially less fortunate Harriet under her wing and had been convincing her to set her sights on men who were in fact socially and financially above her. Knightley was Emma's bachelor brother-in-law, ten years or so older than Emma, a frequent guest in Emma's father's house, and a very stable and important part of her life. She never entertained the

1. Jane Austen, *Emma* (1815) (London: Penguin, 1987), 398.

thought of Knightley as a possible husband for herself, nor as a possibility for Harriet either. Thinking that Harriet had hoped to win Frank Churchill's heart, Emma encouraged her in this pursuit, and Harriet misinterpreted Emma as encouraging her to hope for Mr Knightley. The misunderstanding is straightened out in the scene preceding the one described above.

Consider what might occur to Emma as she hears Harriet's words and "becomes acquainted" with her own heart. Perhaps when Harriet informs Emma of her love for Knightley, Emma has a flash of mental imagery: she pictures Harriet, absolutely radiant, looking lovingly up at Knightley who is standing beside her. Upon having this fantasy Emma might feel vaguely unpleasant; perhaps she feels a little queasy. Or perhaps she has no identifiable physical sensation but instead feels something more complex: vulnerable, as if she were a little girl, alone in the house, wondering why her parents are so late coming home. Then when Harriet offers her reasons for believing that Knightley returns her love perhaps Emma feels even worse: perhaps she feels downright sick or has a memory flashback of when, as a girl, she learned of her mother's death. Or perhaps she merely experiences a little confusion and finds it difficult to engage with Harriet in the way she had been doing a moment before. These are examples of how sensations, emotions, and memory flashbacks might combine with mental imagery to aid Emma's self-discovery. But we could imagine different combinations just as easily. For example, Emma might have no imaginative experience at all; she might simply begin to regard Harriet with contempt, or she might have a sudden impulse to run outside.

Are any of these scenarios unnatural or implausible? Austen implies that in the few minutes of meditation something happened to Emma to enable her to know her own heart or to become aware of her love for Knightley. It is not implausible to suppose that during this time Emma experienced an unpleasant sensation, memory flashbacks, a sudden longing, or a peculiar string of thoughts that occurred while she entertained the thought of Harriet in love with Knightley, and it is not implausible to suppose that the reaction intensified as the evidence supporting the possibility of reciprocation grew stronger. Nor is it unnatural to suppose—though it certainly is not necessary—that it might have as a result of *noticing* that she had an unpleasant sensation or peculiar fantasies accompanying the thought of the two of them together that Emma was able to interpret her own thoughts and feelings. Furthermore it is plausible to suppose, as Austen suggests, that she could not recognize the meaning of, or put an interpretation on, these experiences directly but within a few moments quickly diagnosed them as issuing from, or being understandable in the light of, a judgment that Mr Knightley should marry no one but herself. And surely this judgment, itself, is given sense by supposing Emma to be in love with Knightley.

What is the role of reflection or introspection in self-ascription? An introspective story is natural to tell in Emma's case, and certainly it makes sense to suppose that we often interpret ourselves by making out connections between past-ascribed thoughts, sensations, memories, waking fantasy, and dreams. This raises a question about the role played by these connections and awareness of them in general. Most of the time we do not need to reflect or introspect in order to self-ascribe beliefs or desires, nor do we need to deliberate in order to self-ascribe intentions. But I wish to urge that the practice of treating the subject's expressions of thoughts as in some sense basic data for interpretation is rendered utterly mysterious if we overlook the role that reflection, introspection, and deliberation often play in self-interpretation.

Still, one might not be impressed by Emma's case. After all, love is a complicated matter, and one often takes awhile to identify the complex of emotions and cognitive states and its objects that constitute it or accompany it. One might even misconstrue this network since matters of love are often matters about which we are self-deceived (or so it is said). The problem, though, is that all thoughts are potentially tied up in complex ways with emotions and feelings, bring back memories, invoke fantasies, and are ripe for self-deception. This may be more obviously so for desires and intentions, but it would be true for any intentional state since they depend, for their very nature, on the role they play in psychological explanation, and are necessarily, at least to some extent, intertwined.

2. Crispin Wright has argued that Wittgenstein's "rule-following considerations" are best seen as militating against a certain epistemological story that reductive accounts of meaning impose.[2] Wright resists Kripke's skeptical conclusion that there are no facts about meaning by denying that one's justification for a particular use of an expression must be determined by facts about one's nonintentional phenomenological states. He argues that the challenge posed by Kripke's skeptic to provide a justification for our use of a particular expression can be met by appealing directly to an intention that we had to use an expression in a certain way. Kripke anticipates this response and rejects it as begging the question.[3] The question, according to Kripke, can in fact be turned against my ability to justify knowledge of my own thought contents (intentions) by appeal to nonintentionally described episodes of consciousness and thus the factuality of my own thoughts is called into question. Wright, however, blocks this move straight off on the grounds that it runs headfirst into contradictory first-person facts: in avowing an intention no justification is needed.[4] However, according to Wright, reflection on Wittgenstein's discussion will show that similar worries about justification do seem to threaten the Cartesian picture of how it is one can know one's thoughts directly and be granted authority about them.

---

2. Crispin Wright, "On Making Up One's Mind: Wittgenstein on Intention" (1987), *Rails to Infinity: Essays on Themes from Wittgenstein's* Philosophical Investigations (Cambridge, MA: Harvard University Press, 2001), 116–142; "Moral Values, Projection, and Secondary Qualities" (1988), *Saving the Differences: Essays on the Themes from* Truth and Objectivity (Cambridge, MA: Harvard University Press, 2003), 155–182; "Wittgenstein's Rule-Following Considerations and the Central Project of Theoretical Linguistics" (1989), *Rails to Infinity,* 170–213; "Critical Notice, *Wittgenstein on Meaning,* by Colin McGinn," *Mind 98* (1989): 289–305; "Wittgenstein's Later Philosophy of Mind: Sensation, Privacy, Intention" (1989), *Rails to Infinity,* 291–318. Note that Wright addresses some of my concerns here in "The Problem of Self-Knowledge (I)," *Rails to Infinity,* 319–344, and "Self-Knowledge: The Wittgensteinian Legacy," *Knowing Our Own Minds,* Crispin Wright, Barry C. Smith, Cynthia Macdonald, eds. (Oxford: Oxford University Press, 2000), 13–46.

3. Saul Kripke, *Wittgenstein on Rules and Private Language* (Cambridge, MA: Harvard University Press, 1982), 15ff.

4. Without defending Kripke's meaning-skepticism, I shall just note that I think this move misfires: it seems to be an example of a misconstrual of the role of justification in self-interpretation, which I shall examine in a different context in the next section.

Wright considers §§633–637 from *Philosophical Investigations,* among which, the following, which he takes to be "a graphic expression of the non-inferential character of knowledge of one's own intentions."[5]

> §635 "I was going to say . . . ."—You remember various details. But not even all of them together show your intention. It is as if a snapshot of a scene had been taken, but only a few scattered details of it were to be seen: here a hand, there a bit of face, or a hat—the rest is dark. And now it is as if we knew quite certainly what the whole picture represented. As if I could read the darkness.
>
> §636 These "details" are not irrelevant in the sense in which other circumstances which I can remember equally well are irrelevant . . . .

The proper conclusion to draw from the passages, according to Wright, is that

> the connection between an intention and the act which implements or frustrates it *need not and in general cannot be anticipated by states or processes of consciousness distinct from intending.* To suppose otherwise is to cause the downfall of the notion—as the skeptical argument shows.[6]

If, however, as Wright suggests, Wittgenstein's own comments show that nothing in consciousness *determines* that a certain (content) concept has application this is not to say that nothing in consciousness might *intimate* or *anticipate* the applicability of such a concept; as Wittgenstein suggests, these details are not *irrelevant.*

Wright briefly considers the suggestion that we might look toward phenomenological episodes (flashes of imagery, surges of confidence) as providing inductive evidence for the applicability of a content concept based on past correlations between phenomenological episodes and the

5. Wright, "On Making Up One's Mind," 134.
6. Ibid., 127–128. Emphasis mine.

capacity to deliver the right kind of performance, but he rejects the idea on the grounds that we might with perfect propriety self-ascribe mental states we have never had before. This seems correct: appeal to inductive association is not enough to explain the ability to self-ascribe thoughts that we have never had before, just as it is not enough to explain an ability to recognize sentences that we have never heard before. But Wright concludes from this that we require a different explanation: one that is dissociated from introspection. Having concluded that an explanation of the authority we grant to another's self-ascription cannot be purchased by either a deductive or inductive version of an introspective account, Wright proposes a third explanation:

> So far as I can see, there is only one possible broad direction for such an explanation to take. The authority which our self-ascriptions of meaning, intention, and decision assume is not based on any kind of cognitive advantage, expertise, or achievement. Rather it is, as it were, a concession, unofficially granted to anyone whom one takes seriously as a rational subject. It is, so to speak, such a subject's right to declare what he intends, what he intended, and what satisfies his intentions; and his possession of this right consists in the conferral upon such declarations, other things being equal, of a *constitutive* rather than *descriptive* role.[7]

The view that Wright recommends "would see the [subject's] disposition to make the avowal as *constituting* the state of affairs [that the subject intends that P]. . . ."[8] What a person is prepared to avow "need have no basis"[9]; it "enters primitively into the conditions of identification of what a subject believes, hopes, and intends. . . .[10] [W]hat is being highlighted is a (default) *sufficient* condition for a subject's being in a particular intentional state."[11]

7. Ibid., 137–138.
8. Ibid., 140.
9. Ibid., 138
10. "Wittgenstein's Later Philosophy of Mind," 312.
11. Ibid., 313.

Evaluating the merits of this view will depend on how we construe the claim that the subject's disposition to avow, say, an intention that P *constitutes* the state of affairs that the subject intends that P. A strong version of this thesis—one that denies any achievement is involved in self-ascribing—is untenable; in what follows I shall urge that a weaker version, which I call "constructivist," should be adopted.[12]

Notice that even if we do reject the idea that the basis for self-ascription can be found in the states of consciousness of the self-ascriber, one would still want to claim that the right to self-ascribe in a certain way is, in the general case, *based on something* in the sense that justification conditions for the ascription are relevant. For how else are we to account for the credit we give someone for having psychological insight or the criticism we give him for lacking it?[13] Even if the self-ascription cannot be explained as a kind of *cognitive* achievement, we can still ask what counts as a defeasible or acceptable self-ascription.

---

12. Some of what Wright says, especially the remarks quoted above, seems not to rule out the stronger version, but the gist of his proposal is, I think, more in line with the weaker, "constructivist" line I develop here.

13. Paul Boghossian makes a related point against what he calls "cognitively insubstantial" accounts of self-knowledge, but his conclusion about the inevitability of a cognitive account does not follow. He writes:

> The most important considerations however, against, an insubstantial construal of self-knowledge derives from [the following]; namely, that self-knowledge is both fallible and incomplete. In both the domain of the mental and that of the physical, events may occur of which one remains ignorant; and in both domains, even when one becomes aware of an event's existence, one may yet misconstrue its character, believing it to have a property it does not in fact possess. How is this to be explained? I know of no convincing alternative to the following style of explanation: the difference between getting it right and failing to do so (either through ignorance or through error) is the difference between being in an epistemically favourable position with respect to the subject matter in question—being in a position to garner the relevant evidence or not. To put this point another way, it is only if we understand self-knowledge to be a cognitive achievement that we have any prospect of explaining its admitted shortcomings. "Content and Self-Knowledge," *Philosophical Topics* 17 (1989): 19.

In section III, I shall explain why I think Wright is right to resist the idea that the achievement is a *cognitive* one, or one that tracks an independently existing state of affairs.

Wright claims that although an avowal of intention need have no basis, it is nevertheless criticizable. But "the basis for criticism may only be constituted by states of affairs that were not salient, or even did not exist, at the time of the avowal—par excellence, the subject's subsequent behavior."[14] This claim, however, is puzzling. First, it obscures the point that often we are licensed to challenge ascriptions when they are made without waiting for subsequent nonconformity in behavior; indeed, the fact that subsequent behavior is not *likely* to conform to the ascription may be sufficient for us to challenge it. Second, if this is so, then it cannot be that an avowal of intention *has no basis* or that the authority standardly granted enters *primitively* (nonexplicably) into the conditions of identification of one's intentions. Another way of putting the point: when the ascription is not fulfilled in subsequent action, what grounds do we have for criticizing the individual for *misascribing* an intention as opposed to failing to act in accordance with it?[15] If the former is to be a possible criticism—as it certainly ought to be—then there needs to be something other than the subsequent action which renders the expression or the ascription of the intention appropriate. And these standards, the violation of which allows us to challenge another's self-ascription, ought to be ones to which the subject is held accountable when she self-ascribes. But now an explanation of our practice of granting authority to first-person ascriptions will be parasitic on an explanation why they by and large adhere to (and are not defeasible in the light of) these standards. These standards can be unpacked by looking at the circumstances in which appealing to them would be relevant: when learning takes place, when diagnoses are difficult, and when the ascription is challengeable. What renders a self-ascription acceptable or challengeable depends on whether it is by and large successful in making sense of or fitting into a pattern with the individual's actions, other ascribed thoughts, the environment, and so

---

14. Wright, "Making Up One's Mind," 138.

15. Even if it is implemented in subsequent action, it still might be criticizable; we can imagine a situation in which there are no good grounds for avowing it and its subsequent implementation is, as it were, "accidental."

forth. Wright makes the point himself when he suggests that a positive reason to override a person's avowals would be "the inability of the intentional system so determined to rationalize his behavior satisfyingly, together with the availability of an alternative system . . . which fared better. Or it might be provided by internal constraints of harmony and intelligibility working within the system as a whole."[16] But if the explanatory power of the self-ascription is conceded to be one of the factors which allows it to be accepted or rejected, then the sense in which avowals play a constitutive role needs to be considered carefully.

Wright frames his view by suggesting that judgments about our own mental states fail the *order-of-determination* test.[17] This test concerns the relation between judgments made in cognitively ideal circumstances and the truth of those judgments. By analogy, Wright suggests that our judgments about primary qualities pass the order-of-determination test. Our judgments about shape, for example, can be construed as having a "tracking" epistemology: they *reflect* the truth or standard that is constituted independently of any considerations having to do with us, or our cognitive abilities. By contrast, he suggests that our judgments about secondary qualities fail the test: there is no distance between our judgments and the truth; "truth, for such judgments, is constitutively what we judge to be true when we operate under cognitively ideal conditions."[18]

Wright suggests that at least the a priori presumptive credibility of (what he calls) the "provisional equation":

$$C(\text{Jones}) \rightarrow (\text{Jones intends to } \varnothing \leftrightarrow \text{Jones believes he intends to } \varnothing)$$

can be explained if we read the bi-conditional giving the right-hand side priority and if we construe Jones's belief about his intention as *constitutively determining* it. To construe the (presumptive) credibility of this equation as deriving from an "extension-determining" reading, however, as opposed to one in which its truth consists in the successful

16. Wright, "Wittgenstein's Later Philosophy of Mind," 313.
17. See Wright, "Moral Values, Projection, and Secondary Qualities," and "Wittgenstein's Rule-Following Considerations."
18. Wright, "Wittgenstein's Rule-Following Considerations," 192.

tracking of an independent state of affairs, will only be plausible if there is no better explanation available for its truth or credibility.

Perhaps another way to make the present point is to express the worry that this latter condition is not fulfilled: *another account* seems available of why certain self-ascriptions are acceptable, besides one that sees the subject's best opinions as (presumptively) determining her mental states. Namely, the interpretive story that renders these ascriptions rational.[19] In other words, nothing that has been said so far rules out the idea that interpretation picks out mental states which have a real, causally potent, role in the production of action. Wright should respond, here, that the interpretive story is *not* independent of the subject's best opinions: it relies crucially upon them, thus still making the situation importantly different from concepts with an "extension-reflecting" epistemology, where theoretical (or operational) constraints alone can be considered to be sufficient for establishing the extension of the concepts.

I shall develop the thought that the interpretive story is *not* independent of the subject's best opinions in the next section. It is, I think, important to clarify this, because the thesis that a subject's best opinions (presumptively) determine her intentions may be somewhat misleading if the subject's adherence to "theory," on some reading of that word, is relevant in determining *which* of her opinions are best. I am not sure if, in the end, diagnosing judgments about our own intentional states as failing the order-of-determination test is the best way of characterizing Wright's insight that, in some sense, an individual is able to determine her mental states. It might be more fruitful to investigate why a subject's judgments are required in order for interpretation to get off the ground. In any case, my purpose in this discussion has been to promote the weaker version of the view: one that emphasizes the strong constraints that condition self-ascription, and that emphasizes the individual's ability to see that her self-ascriptions are governed by such constraints. A problem, however, suggests itself. Once we bring in

19. See Jim Edwards, "Best Opinion and Intentional States," *The Philosophical Quarterly* 42 (1992): 21–33.

the role of constraints, or the standards to which one is held responsible when one self-ascribes, have we not deprived ourselves of the obvious fact that, unlike in Emma's case described above, most self-ascriptions seem unmediated and are made without an appeal to standards or to the conditions under which the standards would be met? What about the cases where clearly no such appeal is made? Furthermore, the relevance of standards or constraints reintroduces an obvious role for justification (or evidence). But once the role of justification (or evidence) is reintroduced have we not undermined the presumption of authority granted to first-person reports? In insisting on maintaining a role for standards and justification conditions that govern our ascriptive practices, I have seemingly failed to accommodate the phenomenological fact that we have direct, unmediated access to our thoughts and the presumption of authority granted to first-person reports that gave the Cartesian view its plausibility.

3. Much recent work in philosophy has expressed a worry about relational accounts of content on the grounds that they fail to reflect phenomenological facts; indeed it is commonplace that in many cases we do not need to appeal to anything in order to self-ascribe correctly. But perhaps we should look again at the role that such appeals might play in self-ascription. Davidson sets up the puzzle as follows:

> Because we usually know what we believe (and desire and doubt and intend) without needing or using evidence (even when it is available), our sincere avowals concerning our present states of mind are not subject to the failings of conclusions based on evidence. Thus sincere first[-] person present-tense claims about thoughts, while neither infallible nor incorrigible, have an authority no second[-] or third[-] person claim, or first[-] person other-tense claim, can have. To recognize this fact is not, however, to explain it.[20]

20. Donald Davidson, "Knowing One's Own Mind," (1987), *Subjective, Intersubjective, Objective* (Oxford: Clarendon Press, 2001), 15–38.

This remark is puzzling for two reasons. First, it seems wrong to say that our sincere avowals concerning our present states of mind are not subject to the failings of conclusions based on evidence; they certainly are when they do not make sense.[21] Whether or not justification conditions are appealed to in the act of self-ascribing, the relevance of justification is determined by whether or not the ascription is dubious. This, in general, is when justification for any claim is appropriate. It is a mistake to highlight the mysteriousness of one's relation to one's own mind by pointing to the fact that an appeal to evidence is not, in general, appropriate: justifying a self-ascription is called for when it fails to do its sense-making job. In these cases, unless evidence *can* be adduced to show that an ascription or a self-ascription makes sense in the light of behavior to be explained, phenomenological episodes, and so on, then the ascription ought to be rejected, whether it is a first- or third-person one.

If one is subject to substantial explanatory constraints in self-ascribing, then it is no longer clear how an avowal could be considered primitive. This challenge cannot be met by stipulating that the subject has the conceptual abilities required, that the context is fixed, and then claiming that she *constitutes* the thought in the act of avowing it. For to grant her the conceptual abilities *is,* one might say, to credit her with a substantial achievement. This is most clearly seen for complex applications of mental concepts where we would be inclined to credit a person not merely with competence in using them, but with *psychological insight* when she uses them skillfully.

The "constitutivist" who wishes to retain a strong reading of the thesis might respond by denying that he is talking about the kind of self-ascriptions that manifest insight. He might concede that in cases like Emma's, this account will not do. But he may maintain that for a certain class of second-order thoughts or avowals, there is no context which might render the alleged thought or avowal challengeable. In

21. Note that precisely this point can also be made against Wright's move to block Kripke's skeptic. Recall he argued that Kripke's challenge to produce a justification for avowals of intention flies in the face of first-person facts: that in avowing an intention, no justification is needed. But whether or not one is needed depends on whether the (alleged) intention ascribed meets the intelligibility requirements.

these cases, it might be argued, the subject constitutes her thought in the act of avowing it. Indeed, it is the existence of precisely this class of "basic avowals" or "self-verifying" thoughts that has been invoked to provide the foundation for the so-called "cognitively insubstantial" accounts of self-knowledge.[22] But even if one concedes the existence of this class of thoughts, it needs to be shown how the constitutivist will be able to use it to make general claims about self-knowledge. It is worth noting that presumptive authority and the immediacy of self-ascriptions extends far beyond the class of basic avowals or self-verifying thoughts to all self-ascriptions. In any case, neither the realist nor the antirealist can rely on any such class of judgments, without running straight into the problems that plague Wittgenstein's private linguist. The fact is that all concepts admit of mistaken application; thus all concepts are, in some sense, subject to rules.

I have criticized the strong version of the constitutive thesis for failing to accommodate the sense in which a person can be credited with an achievement if she self-ascribes appropriately; that she ought to be so credited comes out clearly in those cases in which we would be inclined to say she is manifesting psychological insight in her application of mental concepts. But this is not to imply that the achievement can be explained by the subject's theoretical grasp of the rules that govern them.

To ascribe with psychological insight is to react to some situation with the appropriate mental concept, and perhaps to be able, when challenged, to defend the ascription as appropriate. Although we may gesture at them by a compendium of heuristics or generalizations, the appropriate reactions cannot be explained by attributing a prior grasp of rules to the individual. This, then, is another sense in which the ability cannot be explained as a type of cognitive achievement. For

---

22. Burge—in the realist camp—seems to want self-verifying thoughts to play a role in saving first-person authority in the light of the perceived threat to it from an externalist individuation of content. "Sources and Resources of Reason," The John Locke Lectures, Oxford University, 1992–1993. Further, various functionalists who are also in the realist camp adopt an insubstantial cognitive architecture to assure the reliability of self-ascriptions in the light of the perceived threat to authority from an internalist, relational individuation of thought.

such an explanation would involve crediting a person not only with the ability to understand the rule and to perceive the conditions relevant for its application; it would also credit her with the ability to apply the rule in the light of this understanding. In paradigm cases of rule-following, this last condition is satisfied only when a person is able to ascribe to herself the relevant beliefs and intentions. Thus, there is reason to suspect that an attempt to offer an explanation along these lines will be useless, since the very achievement to be explained is presupposed in the type of explanation on offer.[23]

Furthermore, there is no reason to suppose that the rules that govern the extension of our concepts determine a *unique* correct response to a given situation. We should preserve the constitutivist's insight that self-ascribing does not consist in tracking a determinate state of affairs. But—to reiterate the challenge to the constitutivist—it would be a mistake to infer from the fact that we cannot give a cognitive, rule-following explanation of an individual's ability to participate in our psychological practices that justification conditions which render a move in the practice acceptable in the light of some specific challenge, or defeasible if the conditions are not met, are not essential features of any description of that practice. It is consistent, that is, with a *constructivism* about the mental that the various constraints on the application of mental concepts—even if they allow for an element of creativity or indeterminacy—should be included in a description of our psychological practices, just as they are in our artistic practices.[24]

---

23. I describe in more detail one type of regress that ensues in attributing to individuals' cognitive grasp of reasons and norms of rationality in "De-Individualizing Norms of Rationality" (chapter 1).

24. Again, there would seem to be no point anymore in attributing to any individual participant in the practice cognitive grasp of these rules since any *explanation* along these lines would be illusory; the type of explanation on offer presupposes the abilities to be explained. This suggests, *pace* the modern cognitivist trend, reasons for resisting an individualistic psychology for the study of various (social) practices. This point is developed in more general terms in "Playing the Rule-Following Game" (chapter 3), and I discuss it further in relation to Dan Sperber's work in cognitive anthropology in "Investigating Cultures: A Critique of Cognitive Anthropology," *Journal of the Royal Institute for Anthropological Studies* 4, 4 (1998): 669–688.

4. THE FOLLOWING anecdotes might help to crystallize the main points of the discussion. According to legend, when Michelangelo approached a piece of marble to sculpt, he first "discerned" the figure embodied in the stone, and set about "releasing" it from the excess material that trapped it. Mozart (allegedly) reported a relevantly similar experience when he composed: "The whole, though it be long, stands almost complete before my mind, so that I can survey it, like a fine picture or a statue at a glance."[25] These cases nicely illustrate the fact that in what are paradigmatic cases of creation, the phenomenology may be as of perception; from the artist's point of view, it may seem as if he is reporting or describing some independently existing state of affairs. Similarly, in the case of self-ascription, it may seem from the self-ascriber's point of view that he is reporting or describing an independently existing state of affairs.

Early in the film *Amadeus* the young Mozart is introduced to Frederick William II of Prussia playing a piano March written by the court composer, Salieri. When Mozart replays it from memory, he suggests a better ending and introduces variations on the theme: he shows how the music "ought" to go. This example illustrates that though there is a clear sense in which the music is a creation of its composer, it is also true that there are constraints: for example, those of harmony, aesthetic expression, rhythm, and so on, which are not private to him. If they were, Mozart would not have been able to "correct" the composition, and we (the audience) would not have so readily agreed that Mozart's version was superior. Similarly, if in the act of avowing, the self-ascriber is in some sense creating his thoughts, he is nonetheless bound by explanatory constraints of the kind I have attempted to spell out in this chapter.

Self-ascribing, like all conceptual abilities, is a type of skill, and, like experts in all kinds of skills, we are able to self-ascribe without taking time to reflect. Sometimes we act or react so quickly that it does not make sense to suppose that there was anything upon which we base our ascriptions. There may, at times, not be anything in consciousness at all

---

25. Wolfgang A. Mozart, "A Letter," in *The Creative Process*, B. Ghiselin, ed. (New York: New American Library, 1952).

so that self-conscious interpretation would be impossible. Even if there were episodes that anticipate the applicability of content concepts, it would not make sense to say that we study these episodes and choose interpretations *in general.* We simply react in these cases; by and large, we react appropriately. To spell out what renders these reactions appropriate, we can look at various circumstances surrounding the ascription (including a host of counterfactuals about what we might have been aware of, had we reflected—as we did in the case of Emma), which might, under various circumstances, count as justification conditions for these ascriptions.

The ability to react appropriately immediately is an ability that is evinced in various practices that we have mastered. It is manifested linguistically in ones which lean toward the "descriptive" end of the spectrum in which the constraints on what counts as an acceptable move allow less choice on the part of an individual participant. It is manifested, say, musically, in practices which lean toward the "creative" end of the spectrum in which the constraints on what counts as an appropriate response allow more choice, and thus a degree of inventiveness on the part of the individual participant. So, although it is true that in various aspects of our psychological practices we are able to react appropriately immediately, this fact alone should not blind us to the fact that substantial constraints are involved in what constitute *correct* or *acceptable* reactions.

The constructivist thoughts I am teasing out here come out clearly when we look at the cases in which self-ascription takes time and requires reflection as we imagined it to be like for Emma. It rejects the traditional view that there is an object before the mind which is described, but unlike the strong constitutive account, leaves a role for standards and, indeed, evidence to play (this might include an appeal to episodes of consciousness). The constitutivist is right to hold that the conceptual skills involved in self-ascription will not involve tracking a determinate state of affairs and will not be cognitively explicable on this model. Nor, however, will it be correct to say that the abilities are based on nothing. When we want to give an account, say, of musical composition, we will watch how a composer learns his art and how he composes when the process requires thought. When he is able to compose more or less automatically, we will allude to the training and practice that is involved. If an artist reaches fluency—the ability to create spontaneously as if the music, or the figure, were "there," waiting to be described, this does not give us the right to model the creation of the music or sculpture on an "inner theatre" or traditional Cartesian conception, no matter what the phenomenology is like. If Mozart and Michelangelo have experiences in creating that are like those of merely describing what is "already there" this does not mitigate against their roles as creators, nor would we throw up our hands in defeat when trying to say something about the standards which are met, or, indeed, set by their pieces of work. Instead, we will explain the phenomenology by noting that they have reached (or were born with) an amazing stage of proficiency. We may then go on to say in what their *proficiency* consists; not by saying what they cognize, but by outlining the standards that are met (or set) by their reactions. An ability to self-ascribe should be understood similarly as an imaginative, creative skill which is the product of practice and expertise, and, possibly, just plain talent.[26] A person is presumptively (though not infallibly) authoritative in self-ascribing because she is reliable, or skilled, in ascribing thoughts that adhere to certain standards, and because she is reliable or skilled in justifying the ascriptions, when challenged.

I have suggested, though, that this presumption of authority is a compliment paid to one who has mastered any practice; our psychological practices are thought to deliver a special authority to first-person- or self-ascriptions. The special authority we are granted must, I think, inevitably turn on the way in which, and the degree to which, our individual reactions are permitted to shape their appropriateness: and this is the fundamental insight behind the constitutivist's thought. That is, our self-ascriptive practices lean toward the creative end of the spectrum where the constraints on what counts as an acceptable move in the practice allow more inventiveness or choice on the part of

26. Saying this is really just to reiterate the fact that a cognitive explanation of the ability will not be possible, while making clear the role for standards that have been achieved or set by these cases.

an individual participant. In this weak sense, while operating within substantial explanatory constraints, a person might be said *partially* to constitute her thought in the act of avowing it.

One way of understanding this special authority is to notice the degree of flexibility permitted to a self-ascriber who is operating within the constraints: it is here that the notions of "creativity" or "invention" or "decision" seem apt. Consider again the case of Emma. I argued earlier that this is a case of self-discovery. I would also like to show that it is a case of self-invention as well. "It darts though her, with the speed of an arrow, that Knightley must marry no one but herself." Suppose as the thought occurs to her, Emma has a visual fantasy; only this time it is *Emma* and Knightley together at the altar and Knightley is looking lovingly at *her.* Only she is not radiant in reaction to the fantasy, she is immensely disappointed. Or suppose that after the exchange between Harriet and Emma takes place, it darts through her, with the speed of an arrow that Knightley should marry no one but herself; then Emma shrugs, picks up her tatting, changes the subject, and *everything goes on as before.*

Of course, in neither of these cases does Emma realize her love for Knightley; we could not say that in having the thought Emma *believed, judged, decided, or desired* that Knightley should marry no one but herself. Perhaps she tried it on, as it were, but in the scenarios just described, the picture does not appeal.[27] If things happen the way they do in the novel, and Emma not only self-ascribes the belief that Knightley must marry no one but herself, but she goes on to act as if she *accepts* the self-ascription—she acts consistently with it—we might say that she realized her love for Knightley. Self-ascriptions of mental concepts are normative for the self-ascriber of future action and of (re) interpretations of past events; their not fulfilling this role presents a prima facie threat to the acceptability of the self-ascription.[28] Again,

27. I am supposing that we could tell alternative stories making sense of what has happened up until this point: in particular, of (what seems to be) Emma's jealousy. But think of the latitude available to do so: perhaps she does not want her friendship with Knightley threatened, but would not contemplate marrying him either.

28. There is a different kind of case possible—one which is not ruled out by the discussion here—in which Emma is not wrong in self-ascribing the belief that Knightley should

this is perhaps more clearly so for desires and intentions, but it is true for beliefs and its cognates as well.

These constructivist thoughts allow a substantial degree of indeterminacy in what counts as an acceptable deployment of a mental concept. That is—different responses may each meet the various (say, intelligibility) constraints, so that insofar as they do, there is no way of adjudicating between them. The indeterminacy in the constraints will in turn give the subject some choice in what counts as an acceptable justification for the ascription. These indeterminacies in turn provide a self-ascriber with a certain choice in how she is permitted to ascribe at any given moment. But because any one of the appropriate choices might set different substantive (intelligibility) constraints on future actions (and (re)interpretations of past events), in self-ascribing appropriately, a person is also self-determining. We have a special authority about our mental states in the sense that we have some control over whether any particular interpretation of our situation is one we will accept or not: to avow it appropriately is to commit oneself to act consistently with it.[29] Avowing it according to standards governing the appropriateness and acceptability of the ascription is one constraint, but acting in the future in a way which is appropriate vis-à-vis the ascription is another. The justification of thoughts is backward-looking insofar as avowals of them must be intelligible in light of possible descriptions of past behavior and ascriptions of other thoughts. It is forward looking insofar as avowals of

---

marry no one but her. (She acts consistently with it, say, and pursues him.) But she may be wrong that she loves Knightley. Cases of self-deception are certainly possible, or cases in which one correctly ascribes a belief that is mistaken. But in order for this to be possible, many other constraints need to be met, and the point about justification conditions remains intact.

29. To avow it inappropriately, then, might involve a kind of error that is not easily accommodated on the Cartesian model. If love, for example, were simply a feeling that presented itself to me upon mere internal inspection, then I might with perfect propriety tell someone that I love him by correctly identifying the feeling that I have toward him. But if loving someone is conceptually tied to the disposition to make a commitment, then one might wonder about the appropriateness of the avowal when it becomes evident that I am not so disposed. For further reflections, see "Self-Knowledge, Normativity, and Construction" (chapter 15) and "Conceptual Amorphousness, Reasons, and Causes" (chapter 16).

them must be intelligible in light of future behavior. When a first- and third-person ascription compete, and they both satisfy the constraints on, say, intelligibility, the first-person ascription will normally have a privileged status in virtue of the normative constraints upon future action that self-ascriptions place.

Recognizing the choices available to the self-ascriber makes it possible to see how self-ascription can be construed as a kind of creative enterprise, and this, in turn, allows us to make sense of the idea of human freedom. The construction (of the self, as it were) is nevertheless constrained by, among other things, accepted practices of explanation, so our relation to our mind does not threaten to become completely subjective, and the freedom we enjoy is correspondingly limited. The construction of the self is criticizable by objective standards which render its uncriticizability an achievement. Thus, the contrast between the Cartesian and the strong constitutive accounts.

CHAPTER FOURTEEN

Self-Knowledge, Normativity, and Construction

> He tried to look into her face, to find out what she thought, but she was smelling the lilac and the lilies of the valley and did not know herself what she was thinking—what she ought to say or do.
>
> Ivan Goncharov, *Oblomov*

1. Much of modern and contemporary philosophy of mind in the "analytic" tradition has presupposed, since Descartes, what might be called a realist view about the mind and the mental. According to this view there are independently existing, determinate items (states, events, dispositions, or relations) that are the truth-conferrers of our ascriptions of mental predicates.[1] The view is also a cognitivist one insofar as it holds that when we correctly ascribe such a predicate to an individual the correctness consists in the discovery of a determinate fact of the matter about the state the individual is in—a state which is somehow cognized by the ascriber. Disputes have arisen about the nature of the truth-conferrers (e.g., whether they are physical or not) and about the status and the nature of the individual's own authority about the state he is in. A dissenting position in philosophy of mind would have to be handled carefully. It would, most importantly, need to allow for the objectivity of ascriptions of mental predicates at least insofar as it made sense to

1. Henceforth, I shall speak of states or events for ease of exposition. By "independently (or antecedently) existing," I mean states whose existence does not depend on any epistemic interest the subject might take in them.

reject some and accept others on appropriate grounds. Perhaps such a position in the philosophy of mind can be likened in at least one way to what David Wiggins has characterized as a doctrine of "cognitive underdetermination" about moral or practical judgments.[2] In comparing his position of cognitive underdetermination about moral or practical judgments to some things Wittgenstein has said about the philosophy of mathematics, Wiggins suggests that "In the assertibility (or truth) of mathematical statements we see what perhaps we can never see in the assertibility of empirical (such as geographical or historical) statements: the compossibility of objectivity, discovery, *and* invention."[3]

In this chapter I intend to develop the idea that the "compossibility of objectivity, discovery, and invention" is a part of our ordinary (i.e., nonscientific and nontheoretical) understanding of the mental. If this is correct, it is important, since contemporary theories do not make sense of this compossibility; they fail, in particular, to leave room for the inventive aspects of self-ascription.[4]

My strategy involves making judgments about the acceptability or appropriateness of certain ascriptions of mental concepts, so it will be helpful to rely on the description of a possible person and her thoughts made out both in a certain degree of detail and over a significant period of time. I shall be looking at selected details from Goncharov's *Oblomov* and asking my reader to consider some of the scenes concerning the character Olga.[5]

---

2. See David Wiggins, "Truth, Invention and the Meaning of Life"; "A Sensible Subjectivism"; and "Truth, and Truth as Predicated of Moral Judgements," *Needs, Value and Truth* (Oxford: Basil Blackwell, 1987).

3. "Truth, Invention and the Meaning of Life," 130.

4. That they cannot make sense of it comes as no surprise, once it is noted that the goal of so-called "naturalism"—to locate the mental within nature conceived as the realm of law—is ipso facto removing from the mental the first-person point of view or participant perspective that seems so important for retaining the inventive aspect. For this reason it would seem as if any theory of mind that conceives its starting point as the recoil from dualism—(e.g., behaviorism, identity theories, functionalism, and even anomalous monism)—and attempts a full-bodied or modified physicalism, will be unable to account for the inventive or constructive aspects of the mental.

5. Ivan Goncharov, *Oblomov,* (1859), David Magarshack, trans. (London: Penguin, 1954).

2. THE SO-CALLED "Cartesian" model of the mind is one in which discovery plays a role par excellence. This model supposes that one's thoughts, feelings, concerns, needs, values, and principles are exhibited in an "inner theatre" of the mind that is constitutively independent of any epistemic interest the subject might take in it. More problematically, it also supposes that these items are available completely and unmistakably as a result of introspection.

Consider carefully what this picture is committed to. Our experience of the events in which our mentality is allegedly revealed is unmediated and not subject to norms or rules; the events are simply *given* as part of our immediate experience. If our ability to classify them and to recognize them, as classified, is to be infallible this means that our bringing them under concepts—and this would be a matter of applying rules—would not be subject to error. Introducing names for these experiences into our language would presumably involve simple association of the object or event with a name, or "baptism" by ostensive definition. Such classifying or naming, however, would be a private activity, since the experiences classified and named are not accessible to others and the associations cannot be checked by anyone else.

There is a problem with this view. The Cartesian wants the inhabitants of the mind to have an existence that is independent of any epistemic interest taken in them; in this sense he is a realist. He must therefore allow a sufficient gap between what is grasped when the subject "turns his mental eye inward" and his grasping it. One might reasonably require that in order to effect this gap and bring out the true independence of the nature of the objects of the mind it has to be *in some sense* possible for the subject to get it wrong. But the introduction of infallible access thwarts this possibility.

The Cartesian might attempt to dig in his heels and claim that the fallibility associated with our sense-perception of the external world simply fails to apply to the perception of our own minds. Whereas in sense perception the possibility of error is a mark of the independence of the object perceived, he might deny that a viable realism about the mental requires such a possibility. It simply demands that there be a mental item or state that is constitutively independent of the subject's gaze. Success is assured, then, since on this view the "mental eye" and its conceptual machinery functions perfectly.[6]

The opponent of this view must then turn his attention to the idea of perfectly functioning conceptual/perceptual equipment. Indeed, he might plausibly maintain that the very idea of an explanation that posits perfectly functioning machinery is of dubious coherence. For if we posit a mechanism that functions perfectly and cannot go wrong, then we cannot appeal to this mechanism as an *explanation* of the ability. The reason is simply that there would be no way to distinguish any purported explanation using a mechanism that cannot go wrong from a mere description of what would constitute success. As long as it is explanatorily indistinguishable from such a description, there is no reason to posit the mechanism to begin with. And if there is no perfectly functioning mechanism, then the whole idea of objects before the mind that are perceivable by this mechanism is threatened.

The Cartesian model invites us, in effect, to compare the referents of our mental concepts "in their definiteness to objects which are already lying in a drawer and which we then take out."[7] To give this realist aspect of the Cartesian view—the idea that mental items exist in their definiteness independently of any act of identification or endorsement—more chance of success, let us disentangle it from the Cartesian notion of infallibility. The idea that a person might not be aware of what he is thinking is an idea that many people nowadays will be happy to accept. (Many feel this was a discovery of Freud; an idea that along with the Freudian notion of the "unconscious" has not only seeped into our commonsense psychological practices but, in the kindred (though in aspects quite different) form of "tacit" knowledge,

6. See Crispin Wright's discussion of the Cartesian view in "Wittgenstein's Later Philosophy of Mind: Sensation, Privacy, Intention," (1989), *Rails to Infinity: Essays on Themes from Wittgenstein's* Philosophical Investigations (Cambridge, MA: Harvard University Press, 2001), 291–318.

7. Ludwig Wittgenstein, *Philosophical Investigations,* G. E. M. Anscombe, trans. (Oxford: Blackwell, 1953), §193. Wittgenstein uses this metaphor to illuminate the idea that a machine's action seems to be in it from the start (and the metaphor of a machine had been introduced in an attempt to make sense of the idea that an act of meaning can in some sense anticipate reality [§188]).

has permeated contemporary theorizing about language and mind in the cognitive sciences. Some of the minority who remain skeptical about unconscious thoughts have even indicated that the fallibility of first-person ascriptions stands or falls with Freud's technical notion of the unconscious. In my view, both ideas are wrong. Freud's examples of parapraxes in *The Psychopathology of Everyday Life* were convincing because he identified, and put a name to, patterns of action and speech that were candidates for motivated behavior that could be recognized as such by anyone to whom the patterns were pointed out (including the agent himself). I am not familiar enough with literary texts to know when authors started exploring the idea that the intentions and motivations could be discerned without the agent's awareness. But the idea figures commonly in Russia in the works of Dostoyevsky and Goncharov (the latter began writing *Oblomov* in 1849). In France it is evidently to be found in the work of Diderot (*Jacques le Fataliste,* 1796, and it is a major theme in Constant's *Adolphe* [1816], thereby predating the popularization of Freud's work at least in France by a century or more.)

Consider a scene from *Oblomov* in which the fallibility or, in this case, the incompleteness of the subject's own gaze is manifest. Here, the idea that the contents of Olga's mind are apt for "discovery" is especially appropriate. The reader is made aware not only of Olga's words but also of her own thoughts in the form of "inner speech." But some of her thoughts and feelings she is not yet able to recognize: that she is in love with Oblomov, that she was pleased (albeit flustered) by Oblomov's sudden declaration of love, and that she is horrified as he attempts to take it back. Oblomov speaks first, trying to make up with her after rashly declaring his love:

> "Please believe me, the whole thing—I mean, I don't know what made me say it—I couldn't help it," he began gradually growing bolder. "I'd have said it if a thunderbolt had struck me or a stone had crashed on top of me. Nothing in the world could have stopped me. Please, please don't think that I wanted—I'd have given anything a moment later to take back the rash word. . . ."
>
> She walked with her head bowed, sniffing the flowers.

"Please forget it," he went on, "forget it, particularly as it wasn't true. . . ."

"Not true?" she suddenly repeated, drawing herself up and dropping the flowers.

Her eyes opened wide and flashed with surprise.

"How do you mean—not true?" she repeated.

"I mean—well—for God's sake don't be angry with me and forget it. Please, believe me, I was just carried away for a moment—because of the music."

"Only because of the music?"

She turned pale and her eyes grew dim.

"Well," she thought, "everything's all right now. He took back his rash words and there's no need for me to be angry any more! That's excellent—now I needn't worry any more. . . . We can talk and joke as before."

She broke off a twig from a tree absent-mindedly, bit off a leaf, and then at once threw down the twig and the leaf on the path.

"You're not angry with me, are you? You have forgotten, haven't you?" Oblomov said, bending forward to her.

"What was that? What did you ask?" she said nervously, almost with vexation, turning away from him. "I've forgotten everything—I've such a bad memory!"

He fell silent and did not know what to do. He saw her sudden vexation but did not see the cause of it.

"Goodness," she thought, "now everything is all right again. It's just as if that scene had never taken place, thank heaven! Well, all the better. . . . Oh dear, what does it all mean?" [. . .]

"I'm going home," she said suddenly, quickening her steps and turning into another avenue.

There was a lump in her throat. She was afraid she might cry.[8]

Oblomov sees that his attempt to retract his declaration is more distressing to Olga than his original revelation but does not understand

---

8. Goncharov, *Oblomov,* 207–208.

why. She is mollified when he is forced virtually to reproclaim his love, and he is left feeling confused. Only later, in reflecting on the moment when she breaks the lilac sprig, does he come to realize that she loves him. He approaches her again, carrying the lilac sprig, armed with this new knowledge.

"What have you got there?"

"A twig."

"What sort of twig?"

"As you see: it's lilac."

"Where did you get it? There is no lilac here. Which way did you come?"

"It's the same sprig you plucked and threw away."

"Why did you pick it up?"

"Oh, I don't know. I suppose I was glad that—that you threw it away in vexation."

"You're glad I was vexed! That's something new. Why?"

"I won't tell you!"

"Please, do, I beg you."

"Never! Not for anything in the world!"

"I implore you!"

He shook his head.

[. . .]

"What's the matter? Is it something dreadful?" she said, her whole mind concentrated on the question, glancing searchingly at him.

Then gradually realization came to her: the ray of thought and surmise spread to every feature of her face and, suddenly, her whole face lit up with the consciousness of the truth. . . . Just like the sun which, emerging from behind a cloud, sometimes first lights up one bush, then another, then the roof of a house and, suddenly, floods a whole landscape with light. She knew what Oblomov's thought was.

"No, no," Oblomov kept repeating. "I could never say it. It's no use your asking."

"I'm not asking you," she replied indifferently.

"Aren't you? But just now—"

"Let's go home," she said seriously, without listening to him. "Auntie is waiting."[9]

After this scene, Olga goes home and immediately begins acting like a woman in love; I shall discuss this transformation shortly. For the present all we need notice is that these passages illustrate nicely the sense in which "discovery" is an apt description of what sometimes happens in self- and other-ascriptions. It also illustrates nicely the sense in which these ascriptions might qualify as "objective." Whatever pattern of thought and behavior is supposed to indicate a person's mental states, it is often identifiable by others. In this case Oblomov is the first to identify some of Olga's thoughts and feelings. When Olga finally comes to see them, her recognition results from inference or a chain of reasoning: in this case via her realization of Oblomov's thoughts about the significance of her behavior.

3. One of the problems with the Cartesian model is its failure to leave room for a requisite sense of objectivity. Another is its failure to accommodate the intuition that at least for many mental states (paradigmatically ones involving "propositional attitude" concepts) the criticism we incur when we misascribe results from a kind of explanatory failure. Indeed, what generally defeats a self-ascription is its failure to fit into a rationalizing story. According to the perceptual model, defeat is rather a matter of failing to track the private items that exist in the mind's eye. Of course, even on this Cartesian model what is tracked may be—contingently—(part of) an explanatory project. But later philosophy of mind has accorded mental concepts (especially those apt to play a role in reason-explanation) with more than a merely contingent explanatory role: the intuition—which forced those attracted to type physicalism to withdraw to token physicalism—is that propositional attitude concepts in particular (and hence the emotional states that presuppose them) are—constitutively—explanatory concepts.

9. Ibid., 216, 217.

Functionalism—the dominant position in contemporary philosophy of mind—seems to avoid both objections to the Cartesian model and yet retain the sense in which self-knowledge is comfortably seen as a matter of discovery. This philosophical position holds that when we ascribe a mental concept to an individual, it refers to a state of the person that has appropriate causal connections to sensory input, behavioral output, and other internal states.

According to this doctrine, success in ascription is a matter of tracking or homing in upon those states with the appropriate causal specification. Functionalists may still maintain that the point of mental-concept ascription is to render intelligible the one to whom the concepts are ascribed, as long as it is the causal role that is doing the explanatory work. This picture seems to be consistent with ideas exemplified in the scenes from *Oblomov* at least insofar as ascription is tantamount to the discovery of a pattern. It is also consistent with the possibility that the agent is not best placed to notice this pattern (as causal role).

Problems arise, however, when one reflects on what exactly is doing the explanatory work. Although I will not argue for this here, I think it is doubtful whether functionalists can consistently maintain that the explanatory project is a *rationalizing* one or that mental concepts explain in virtue of the way in which "things are made intelligible by being revealed to be, or to approximate to being, as they rationally ought to be." They will not succeed in maintaining this, at least, if such a rationalizing account is "to be contrasted with a style of explanation in which one makes things intelligible by representing their coming into being as a particular instance of how things generally tend to happen."[10]

If, as I believe, these styles of explanation are indeed different, then a question arises about functionalism's relation to our ordinary, commonsense ascriptive practices where these are understood as making fundamental use of *rationalizing* explanations.[11] Functionalism, as

10. John McDowell, "Functionalism and Anomalous Monism," (1985), *Mind, Value, and Reality* (Cambridge, MA: Harvard University Press, 1998), 328.

11. See "Why Reasons May Not Be Causes" (chapter 5).

originally conceived, was a thesis about the meaning of mental predicates, presumably those used within our ordinary, commonsense, psychological practices. As such, its viability as a theory of mind would depend upon whether the concepts ascribed within these practices do, in effect, track internal, functionally individuated (physically realized), causally efficacious states or events.

My suspicion is that they do not. I mention this difference in explanatory patterns—causal on the one hand, and rationalizing on the other—because I suspect it will be of utmost importance. It is arguably the rationalizing pattern and not a causal-explanatory one that allows the reintroduction of an inventive aspect to the role of self-ascriptions.

4. I SHALL NOW consider what is compelling about the idea that a person has some inventive or creative role to play in respect to his mental life. I have discussed how it makes sense to say that Olga's realization of her love for Oblomov came as a discovery (one that Oblomov had made before her). But to call it a "discovery" would only be partly correct. A few hours after Olga realizes her feeling for Oblomov, she becomes transformed.

> He waited nervously and with trepidation for Olga to come down to dinner, wondering what she would say, how she would speak, and how she would look at him. . . . She came down—and he could not help admiring her; he hardly recognized her. Her face was different, even her voice was not the same. The young, naïve, almost childish smile not once appeared on her lips; she did not once look at him with wide-open eyes questioningly or puzzled or with good-natured curiosity, as though she had nothing more to ask, find out, or be surprised at. Her eyes did not follow him as before. She looked at him as though she had studied him thoroughly, and, finally, as though he were nothing to her, no more than the baron—in short, he felt as though he had not seen her for a whole year during which she had grown into a woman.[12]

12. Goncharov, *Oblomov,* 222–223.

Olga is transformed from someone who was (arguably) in love into someone who now acts in self-conscious awareness of her love, or rather in accordance with her own conception of how a woman in love should act. Might not this passage suggest that there is something right about the idea that the nature of the love she has "discovered" is changed as a result of these subsequent actions, and hence as a result of this self-awareness? The idea would be not merely that her love for Oblomov causes her transformation into a "woman" or even that her *awareness* of it does, but rather that her awareness and her endorsement of it somehow affect the love or the shape of the love itself. They play a role in a more complex "rationalizing project" that involves her own conception of how a woman in love should act. This explanation of her behavior (that she is in love with Oblomov), its endorsement by Olga, and its role in an ongoing narrative give shape to, or articulate, a pattern or a possibility which in turn (retrospectively, as it were) supports the original description of Olga as a woman in love.

Charles Taylor has, in a series of articles, attempted to argue for the idea (which he credits to Heidegger) that a person's self-conception partly constitutes the mental state he is in.[13] He argues that much of what we think, feel, and value is not the result of our being moved by forces like gravity or electromagnetism. Our desires and aspirations are given formulation in words or images; they cannot but be articulated or interpreted by us somehow. But these articulations

> are not simply descriptions, if we mean by this characterizations of a fully independent object, that is, an object which is altered neither in what it is, nor in the degree or manner of its evidence to us by the description. In this way my characterization of this table as brown, or this line of mountains as jagged, is a simple description.
>
> On the contrary, articulations are attempts to formulate what is initially inchoate, or confused, or badly formulated. But this

13. See especially "What is Human Agency?" (1977), "Self-Interpreting Animals," and "The Concept of a Person" (1981), in Charles Taylor, *Human Agency and Language: Philosophical Papers* (Cambridge: Cambridge University Press, 1985).

> kind of formulation or reformulation does not leave its object unchanged. To give a certain articulation is to shape our sense of what we desire or what we hold important in a certain way.[14]

Taylor gives his own example of what he means, but the point can be developed by staying with the character of Olga and by noting how Olga's endorsement of herself as a woman in love gives shape to, or articulates what—though it amounted to a "discovery"— had been inchoate or confused before.

An analogy might be helpful. Think about a duck-rabbit design, which, although ambiguous between being either the head of a duck or a head of a rabbit, is arguably not the head of a cow or pig. Now, imagine that when the figure is drawn with more detail (a body is added) it becomes a duck and not a rabbit. The analogy would be that Olga's pattern of behavior before her reflections was in certain ways indeterminate (though certain interpretations of her behavior could be ruled out) just as the duck-rabbit design is ambiguous (though certain descriptions of the design can be dismissed). After her reflections and her endorsement of one pattern (on our analogy, she recognizes the pattern as a duck), she behaves in a way that is consistent with that recognition. Her endorsement of it (as a duck) and her subsequent behavior allow the pattern to develop in such a way (say, it develops a beak, webbed feet, feathers, etc.) that renders the other interpretation no longer viable.

This idea can be spelled out in more detail when we consider what happens to Olga later in the novel. Her relation with Oblomov has come to a painful end, and she has slowly started to enjoy, and depend more and more upon, the company of her old friend, Stolz. Stolz falls in love with Olga and she is confused about her feelings for him.

> If she loved Stolz, then what was her first love? Flirtation, frivolity, or worse? She blushed with shame and turned hot at this thought. She would never accuse herself of that. But if that was her first

14. "What is Human Agency," 36.

> pure love, what were her relations to Stolz? Again play, deception, subtle calculation, to entice him into marriage so as to cover up the frivolity of her conduct? She turned cold and pale at the very thought of it. But if it was not play, or deception or calculation—so . . . was it love again? But such a supposition made her feel utterly at a loss: a second love—eight or seven months after the first! Who would believe her? How could she mention it without causing surprise, perhaps—contempt! She dared not think of it. She had no right. She ransacked her memory: there was nothing there about a second love. She recalled the authoritative opinions of her aunts, old maids, all sorts of clever people, and, finally, writers, "philosophers of love"—and on all sides she heard the inescapable verdict: "A woman loves truly only once."[15]

Olga concludes that what she feels for Stolz must only be a sisterly love. Stolz confronts her about her baffling behavior, and she is eventually forced to confess that she had been in love with Oblomov, and she tells him the whole story of their courtship. When she then shows Stolz a letter Oblomov had written to her very early in their relationship, Stolz uses it to interpret Olga's past feelings rather differently.

> "Listen," he said, and he read: "'Your present *I love you* is not real love, but the love you will feel in the future [. . .] You have *made a mistake*' (Stolz read, emphasizing the words) 'the man before you is not the one you have been expecting and dreaming of. Wait—he will come, and then you will come to your senses and you will feel vexed and ashamed of your mistake.'" [. . .] "You see how true it is," he said. "You were vexed and ashamed of—your mistake. There is nothing to add to this. He was right and you did not believe him—that is all your guilt amounts to." [. . .]
>
> "I did not believe him. I thought one's heart could not be mistaken."

15. Goncharov, *Oblomov,* 400–401.

> "Yes, it can, and sometimes very disastrously! But with you it never went as far as the heart," he added. "It was imagination and vanity on one side, and weakness on the other."[16]

When Olga comes to accept this new interpretation of her feelings for Oblomov (albeit mistakenly, I would judge), her love for Stolz becomes a possibility for her in a way that it could not have been without this change of self-conception. She could not rationally hold that she was in love with Oblomov, that a woman only loves once and that she now is in love with Stolz. With Stolz's encouragement she gives up the idea that she had been in love with Oblomov. Once her self-conception or "practical identity" has been made consistent, she is able to reinterpret her feelings for Stolz as more than mere sisterly love and thereafter allows herself to act freely upon this new conception.[17] The romantic love for Stolz thus takes shape. It is presumably this sort of phenomenon that leads Taylor to claim that

> [w]e can say therefore that our self-interpretations are partly constitutive of our experience. For an altered description of our motivation can be inseparable from a change in this motivation. But to assert this connection is not to put forward a causal hypothesis: it is not to say that we alter our descriptions and then *as a result* our experience of our predicament alters. Rather it is that certain modes of experience of our predicament are not possible without certain self-descriptions.[18]

---

16. Ibid., 411–412. Oblomov wrote the letter that Stolz refers to out of a mixture of cowardice and vanity: partly in an attempt to derail the impending complication that such a relationship would bring to his life, and partly to witness Olga's distress as she reads the letter. His claim that Olga does not really love him is one I would reject, but it is an interpretation with which Stolz can reassure Olga.

17. The term "practical identity" is from Christine Korsgaard, *The Sources of Normativity* (Cambridge: Cambridge University Press, 1996).

18. Taylor, "What is Human Agency?" 37.

Compare:

> Our contingent practical identities are, to some extent, given to us—by our cultures, by our societies and their role structures, by the accidents of birth, and by

5. Richard Moran has recently argued, pace Taylor, that the sense in which a person's self-conception affects his emotions or other first-order mental states is not a logical, individuative, or constitutive one: it is causal.[19] If his argument is sound, then it might undermine the constructivist aspects of self-knowledge I am attempting to bring into view and lend plausibility instead to the idea that our mental concepts refer to states that play a complex causal role.

There is a constant slide, Moran argues, between two different stances we might take. On the one hand, we might take what he calls a "theoretical stance" toward our mental states qua independently existing objects and describe or track them. On the other, we might take a practical stance toward them and make a restricted, "indirect" decision about what to believe or what to intend. This latter, practical question about what to believe is "transparent to," or answered in the same way as, our theoretical questions about (what is true in) the world.

According to Moran, the rationality of agents has a dual aspect: it ensures that a person's beliefs will aim at the truth, and it ensures that a person's second-order beliefs about his own mental states will affect his first-order beliefs. For example, if the self-interpreter notices an inconsistency or a contradiction in his first-order beliefs his theoretical question about what he believes "involves reasoning guided by the question of what is true about the object of belief."[20] The idea, presumably, is that the theoretical question about what I believe will transform itself into a practical question about what *to* believe, since the observation that one's belief is false is, at least prima facie, sufficient to destroy it. Moran insists that this relationship between second-order and first-order beliefs is not to be construed as a constitutive or logical relationship. It is simply a matter of the tendency of theoretical questions to transform

our natural abilities—but it is also clear that we enter into their construction. And this means that the desires and impulses associated with them do not just arise in us. When we adopt (or come to wholeheartedly inhabit) a conception of practical identity, we also adopt a way of life and a set of projects, and the new desires which this brings in its wake. (Korsgaard, *The Sources of Normativity,* 239)

19. Richard Moran, "Making Up Your Mind," *Ratio* 1 (1988): 148.

20. Ibid., 148.

themselves into practical ones. He concludes that "self-understanding and self-change *can* be understood in a way that maintains the logical independence of interpretation and its object."[21]

I think Moran avoids, however, the crucial issue in characterizing one of the stances I might take as *practical* instead of *normative.* On the latter conception, it will be much more difficult to make out a contrast between two different stances, since the normative cuts across the distinction between the theoretical and the practical. There would be some truth, for example, in saying that theoretical questions about what I believe can transform themselves into normative questions about what I should believe, in a way that is transparent to questions about what is true in the world. But questions about what I believe also are influenced by what would explain (rationalize) my actions (both past and anticipated present ones), what would best cohere with the other beliefs that I hold or have held, as well as my principles, long-term projects, and so forth.

In order to make out his case that beliefs are logically independent of the subject's gaze, Moran suggests that we focus on a person's *past* beliefs and his present interpretations of them, because practical questions do not apply to these and in such cases, the theoretical question about what I believed will not be influenced by practical questions about truth (and about what *to* believe). The idea is that the theoretical identification of a past belief will not transform itself into the practical one about what *to* believe, since the practical question is now out of date.

Questions, however, about what would explain my actions, what would best cohere with other beliefs, and so forth, might well influence my identification of a past belief as much as they will influence my identification of a present one. Indeed, because of the pervasiveness of these normative criteria on belief identification, it is simply not clear that there is a viable distinction between a theoretical and a normative stance I might adopt toward my own mental states. (Notice how this point is suggested in the passage from *Oblomov* cited as the epigram to this paper: Olga did not know herself what she thought—what she

21. Ibid., 149.

ought to say or do.) Moran might be right in saying that I can ascribe to myself a past but not a present belief that I know to be false. Nonetheless, because of the other normative constraints (besides the aim for truth) on belief identification, there is plenty of scope to reintroduce the idea that a person's self-conception plays a constitutive, and not merely causal, role in *shaping* (and not merely in describing) his first-order mental states.

In order to pursue this idea further, consider the obligations or entitlements that self-ascription, or indeed any epistemic claim, imposes. A particular ascription, for example, will commit the ascriber to a certain range of justifications he might give for it, if challenged. It will commit him to a range of considerations that would count against it or would follow from it, and included here, of course, would be a certain range of actions. Should a sufficient number of these commitments fail to obtain or to be endorsed, this puts increased pressure on the self-ascriber to withdraw the original ascription. To take a simple example, suppose that my choice of restaurants might be explained by either the quality of the food or the location. My accepting the latter as a reason ought to affect my attitude toward the suggestion that I might find equally good food elsewhere. I might, of course, be wrong about why I chose the restaurant (and the alacrity with which I agree to go elsewhere might suggest a reason to suspect that I was). In this case, I ought to reevaluate my reason for choosing this restaurant. My continued acceptance of location as a reason even if I agree without hesitation to go elsewhere creates a tension which needs to be eased either by introducing other reasons for acting into the picture (e.g., I recognize that my companion wants very badly to dine elsewhere) or by my construing my decision to go elsewhere as one that fails to reflect my preference. This is similar to the unstable position Olga found herself in when trying to sort out her feelings for Stolz. My own understanding of my mental attitudes carries with it rational constraints on my future choices, decisions, actions, explanations, criticisms, and justifications. This is true for reasons or attitudes that I attribute to myself as a result of reflection or interpretation; it is also true for immediately ascribed expressions or avowals.

Intentional actions, propositional attitudes, and affective states that presuppose them are identified as such by their role in a pattern of other thoughts and actions. I would like to suggest that if the pattern is a rationalizing one, and thus explanatory in the sense that "things are made intelligible by being revealed to be, or to approximate being, as they rationally ought to be," then at least many patterns will be "open-ended" and lend themselves to further, and perhaps different, interpretations. Consider how John Wisdom characterizes the idea:

> Suppose two people are speaking of two characters in a story which both have read or of two friends which both have known, and one says "really she hated him," and the other says "she didn't, she loved him." Then the first may have noticed what the other has not although he knows no incident in the lives of the people they are talking about which the other doesn't know too, and the second speaker may say "she didn't, she loved him" because he hasn't noticed what the first noticed, although he can remember every incident the first can remember.[22]

Like an aesthetic dispute about, say, the beauty of an object, or a legal dispute about whether reasonable care has been exercised, reasons for or against a certain judgment can be adduced. But in cases such as these,

> we notice that the process of argument is not a *chain* of demonstrative reasoning. It is a presenting and representing of those features of the case which *severally co-operate* in favour of the conclusion, in favour of saying what the reasoner wishes said, in favour of calling the situation by the name by which he wishes to call it. The reasons are like the legs of a chair not the links of a chain.[23]

This is plausibly the case when we make evaluative judgments in matters of ethics or in practical deliberation about what it would be rational to do as well. Wiggins makes a similar point in a passage in which

---

22. John Wisdom, "Gods," (1951), *Logic and Language* (Oxford: Blackwell, 1963), 191–192.
23. Ibid., 195.

he discusses an idea that can be salvaged from a naïve noncognitivism in ethics and imported into a more sophisticated doctrine of cognitive underdetermination:

> . . . not all the claims of all rational concerns or even of all moral concerns (that the world *be* thus or so) need be actually reconcilable. When we judge that this is what we must do now, or that that is what we'd better do, or that our life must now take one direction rather than another direction, we are not fitting truths (or even probabilities) into a pattern where a discrepancy proves that we have mistaken a falsehood for a truth. Often we have to make a practical choice that another rational agent might understand through and through, not fault or even disagree with, but . . . make differently himself. . . .[24]

It seems to me that the freedom alluded to here with respect to our practical choices figures as well as a feature of our interpretive practices. I suggest that it is arguably indeterminate at the time she broke the lilac whether Olga loved Oblomov just as it was arguably indeterminate whether, at the time she was confronted by Stolz, she had sisterly or romantic feelings for *him.* (I think it is indeterminate whether Constant's Adolphe, in seducing Ellénore, was really in love or was rather simply carried away by the intensity of what had been a game.) Their "self-takes" play a role in "articulating" what had been indeterminate before.[25]

---

24. Wiggins, "Truth, Invention and the Meaning of Life," 126.

25. Or, in cases of self-deception, confusing, or muddling what had been inchoate before. In these cases there are at least two strands of thought and action patterns manifested. One is the pattern that belies the agent's self-conception and uncovers her ignorance about her own mind. The other is the pattern—often of denials, of protestations, of avoidance—that is a straightforwardly rationalizable outcome of this self-conception. Adolphe, who is self-deceived about the obstacles to his worldly success, is not merely wrong to blame his relationship with Ellénore. His false conception about their life together feeds into a whole pattern of behavior leading to a tragedy that is itself only rendered comprehensible by this conception.

The case I am making works for states that are inchoate, confused, or multiply interpretable. Not all mental ascriptions or avowals fall into this category. This may be because the subsequent commitments to which an ascription or avowal is answerable have been largely fulfilled, or because it involves relatively little by way of such commitments, like my expression of the desire to have a glass of wine after I have finished work for the day. Other of my desires and intentions—to develop my singing voice, to complete an edging of Bucks Point lace, to expend no more and no less than a reasonable amount of effort doing philosophy—are more complex. They involve commitments extending well into the future that will come into contact and conflict with other intentions, short-term desires, and perhaps some principles or values too. It is true that in considering the application of a concept like love, I am considering complex and pivotal (or central) patterns of action, running from and to numerous other subpatterns. But while philosophy of mind's discussions tend to take simple, discrete actions (like raising one's arm to signal) as its paradigm case, it is committed to explaining the mental states that figure in an explanation of the projects, plans, commitments, and so forth that constitute a person's life as well.

That it is a *person*'s life is important here. Patterns of animal behavior can be identified in and rationalized by intentional psychological terms.[26] But although animals can act in accordance with some rational

---

26. Including, perhaps, the "differential response dispositions" shown by thermometers. Taylor ("What is Human Agency," 28) suggests parenthetically that Camus's Mersault might be an example of someone who fails one test of personhood insofar as he lacks the ability to "deploy a language of evaluative contrasts ranging over desires" (ibid., 23). Consider another character from Oblomov. Agafya Matveyevna Pshenitzyn, the woman whose elbows entrance Oblomov and eventually capture his heart, is described by the narrator as someone barely capable of self-reflective awareness.

> Had she been asked if she loved him, she would again have smiled and said yes, but she would have given the same reply when Oblomov had lived no more than a week at her house. (*Oblomov*, 374)

> He was a gentleman: he dazzled, he scintillated! And, besides, he was so kind; he walked so softly, his movements were so exquisite; if he touched her hand, it was like velvet, and whenever her husband had touched her, it was like a blow! And he

norms (and this might suffice to ascribe intentional states to them) they lack the meta-ability to understand what the norms commit them to. This will involve an ability to see ways in which a pattern might continue consistently with certain identifications but inconsistently with others. And this ability to recognize patterns and to act in accordance with them because they have been endorsed will introduce a complexity to the patterns that would have been inconceivable for non-self-reflective beings. If what I have been arguing here is correct, part of this complexity will involve a kind of self-construction. It is the role of the (whole) person in this construction—the understanding of the commitments and obligations of a rational agent—that seems ill accommodated by causal, reductionist accounts.

6. I HAVE SUGGESTED elsewhere that constructivist considerations about the mental will go some of the way toward explaining the authority of first-person applications of mental concepts and the asymmetry between first- and third-person ascriptions.[27] Such considerations, in order to be plausible, will also grant that the choices available in interpretation are not free or unconstrained—anymore than are the choices available in musical composition. W. C. Kneale has argued that the important contrast for the constructivist is one that emphasizes the difference between geographical or historical claims on the one hand, and

---

> looked and talked so gently, with such kindness. . . . She did not think all these things, nor was she consciously aware of it all, but if anyone had tried to analyse and explain the impression made on her mind by Oblomov's coming into her life, he would not be able to give any other explanation. (Ibid., 375)

> Agafya Matveyevna herself was not only incapable of flirting with Oblomov and revealing to him by some sigh what was going on inside her, but, as has already been said, she was never aware of it or understood it herself. . . . Mrs Pshenitzyn's feeling, so normal, natural, disinterested, remained a mystery to Oblomov, to the people around her, and to herself. (Ibid., 376)

Her brother even characterizes her as an animal:

> "She can't be expected to look after her interests, can she? A cow—that's what she is, a blamed cow: hit her or hug her, she goes on grinning like a horse at a nosebag-ful of oats." (Ibid., 357)

27. See "Some Constructivist Thoughts About Self-Knowledge," (chapter 13).

mathematical, and some ethical, practical, and aesthetic claims that are likened to artistic creation on the other.[28] No plausible use of "invention" in this context would suppose that it means being capable of creating possibilities from nothing as some—incoherently, he thinks—believe God capable of. "An artist can do no more than select an interesting possibility."[29] As long as the distinction between say, geographical and mathematical claims is kept in mind (e.g., that America existed before the first men landed there but the infinitesimal calculus did not exist before it was first formulated), then, Kneale argues, the terms "invention" and "finding" may both be apt, since there is no relevant difference between making-with-the-mind and finding-with-the-mind. The contrast between the cases is rather (partly) between what we find with the sense organs and what we find with the mind.

If it is conceded that both the Cartesian and the functionalist suppose that it is something on analogy with a sense organ that "finds" or "discovers" the denizens of the mind—the "mind's eye" on the first model, and an internal scanner on the second—then the contrast is one I can adopt. I am arguing for a rejection of this mode of discovery, and am plumping instead for the discovery or selection of something akin to an "interesting possibility." This would allow us to begin making sense, then, of the "compossibility" of objectivity, discovery, and invention in the area of psychological discourse.

28. W. C. Kneale, "The Idea of Invention," *Proceedings of the British Academy* 39 (1955): 85–108.

29. Ibid., 101.

CHAPTER FIFTEEN

Speaking One's Mind

Dorit Bar-On's *Speaking My Mind: Expression and Self-Knowledge* is about avowals, or a particular subset of what Ryle called "unstudied talk" about oneself, which employs (in Ryle's words) "explicit interest phrases" like "I want," "I hope," "I intend," "I dislike," "I am depressed," "I wonder," "I guess," and "I feel hungry."[1] One striking feature of avowals, noted by Ryle, is that they seem to enjoy a special kind of security from epistemic assessment or criticism. "How do you know?" or "I think you must be mistaken" or "You have been careless in your observations" do not make sense as rejoinders to avowals. Yet they are plausible rejoinders to their "semantic cousins": e.g., nonavowal ascriptions such as "she is in pain" (said by you about me) or "JT is in pain" (said by either of us about me) or "I am in pain" said by me on the basis of observing my frowning face in the mirror or on the basis of observing myself limp, or (perhaps) even "I was in pain" said by me about a moment in the past. What

1. Dorit Bar-On, *Speaking My Mind: Expression and Self-Knowledge* (Oxford: Clarendon Press, 2004). Gilbert Ryle, *The Concept of Mind* (1949) (London and New York: Routledge, 2009), 164.

protects the avowal from epistemic assessment or criticism? Here is Ryle's answer:

> If the avowal ["I feel depressed"] is to do its job, it must be said in a depressed tone of voice; it must be blurted out to a sympathizer, not reported to an investigator. Avowing "I feel depressed" is doing one of the things, namely one of the conversational things, that depression is the mood to do. It is not a piece of scientific premiss-providing, but a piece of conversational moping. That is why, if we are suspicious, we do not ask, "Fact or fiction?" or "True or false?," "Reliable or unreliable?" but "Sincere or shammed?" The conversational avowal of moods requires not acumen, but openness. It comes from the heart, not from the head. It is not discovery, but voluntary non-concealment.[2]

And later, noting that while people can be reticent or hypocritical, Ryle urges that nonetheless reticence and hypocrisy are learned behaviors.

> In a certain natural sense of "natural" the natural thing to do is to speak one's mind, and the sophisticated thing to do is to refrain from doing this, or even to pretend to do this, when one is not really doing so.[3]

Bar-On's thought-provoking book—though it does not include these passages from Ryle—develops the idea (inherited from Wittgenstein) that in avowing, we speak our minds.

> . . . avowing "I am hoping it doesn't rain" is like saying or thinking "It doesn't rain" hopefully. On the present proposal, the point of avowing an intentional state is not to provide a descriptive report of it, but rather to share it, or air it, or give it voice, or just to "vent" it.[4]

---

2. Ryle, *The Concept of Mind*, 87.
3. Ibid., 162.
4. Bar-On, *Speaking My Mind,* 220.

However,

> [u]nlike the person who grunts, or smiles, [the avower] is *speaking* her mind—that is, she is using articulate verbal means to express her state of mind.[5]

It is because avowers use articulate verbal means to express their states of mind that Bar-On thinks her "neo-expressivist" treatment of avowals will also accommodate—besides "epistemic asymmetry"—the "semantic continuity" that allows avowals to share the truth-conditions and participate in logical inferences with other (e.g., first- and third-person) nonavowed ascriptions that identify an individual and attribute to her a mental condition. Bar-On develops her neo-expressivist account in order to explain avowals' immunity from various kinds of errors, as well as their presumptive truth. Her account goes beyond a "simple" expressivism which assimilates avowals to natural expressions. Bar-On's neo-expressivism holds that

> avowals are similar to natural expressions in terms of the *acts* that avowing subjects perform; but their products are different. Avowals as products have semantic structure. Of the product of an act of avowing (i.e., a self-ascription of a present mental state) we can say that it expresses [in the semantic sense] the proposition or thought that the subject is in a certain state. For the avowal-as-product is a conventional representation of that thought. And a component of it—viz., the mental term—conventionally represents the relevant mental state. Not so in the case of natural expressions.[6]

The accommodation of "epistemic security" together with the preservation of "semantic continuity" are the main desiderata of the account, but the author also hopes to show that avowals' special security is fully consistent with a robustly realist view about the nature

5. Ibid., 262.
6. Ibid., 255.

of mental states and with the idea that the avower has privileged and genuine knowledge about the states she avows. (It is this combination of views that I find most puzzling, as I shall explain at the end.)

Ryle considers the phenomenon of avowals or "unstudied talk" as part of the wider topic of self-knowledge. (He argues that the self-ascriber and other-ascriber are in a similar position when it comes to knowing or finding out about the former's moods, inclinations, and frames of mind.) But insofar as Bar-On is interested in the topic of self-knowledge, it is restricted here to her account of the special security of avowals. Although she admits that authorship or "construction," endorsement, commitment, and the like, can play a role in the larger topic of self-knowledge, she maintains that these features cannot help us understand the epistemic protection (nor presumably the "aptness" for knowledge) enjoyed by avowals (phenomenal as well as intentional)—a security that does not protect the avowals against charges of error tout court—but only of a specific kind of epistemic error.

Bar-On takes her nemesis to be the same as that of Wittgenstein (and, we might, of Ryle): the Cartesian "inner theatre" model of mental states that would diagnose the special security of avowals as a form of inner perception. In one of the best discussions I have read of (one aspect of) Wittgenstein's polemic against the possibility of a private sensation language, Bar-On sets the stage for a different, not only non-Cartesian, but *nonepistemic* account of avowals' special security.

There are two kinds of security that Bar-On alleges for avowals. One who avows is protected from criticism for failing to identify the referent of "I" because (to use Wittgenstein's expression) the possibility of error has not been provided for.[7]

According to Bar-On, however, one is also protected from criticism (in the case of genuine avowals) for having mistaken the kind of attitude one avows (desire, need, want, hope, etc.) as well as the intentional

7. Wittgenstein said "No error is possible because the move which we might be inclined to think of as an error, a 'bad move,' is no move of the game at all. . . . It is impossible that in making the statement 'I have toothache' I should have mistaken another person for myself, as it is to moan with pain by mistake, having mistaken someone else for me." Ludwig Wittgenstein, *The Blue and Brown Books* (1958) (Oxford: Blackwell, 1975), 67.

object or content of the avowal (when there is one—say, that it is a meal that I desire rather than a walk in the country). In avowing, one is protected from epistemic criticism not only for failing to refer but also for failing to identify the attitude, the intentional object, or the propositional content ("intentional content"), because one's act of avowing does not purport to be the discovery or identification of something which requires any epistemic effort on the part of the avower.

Bar-On is explicit that the kinds of epistemic security that attend avowals do not protect it against all kinds of error. On her account a person can make mistakes ("expressive mistakes") both about the kind of state she is in and also about its intentional content. But this will not be because 1) the world has let her down (i.e., she is not subject to "brute error") or 2) some recognitional or identificatory capacity is deficient. No epistemic method has failed her; or, as she sometimes puts it, a self-ascription does not rest, epistemically, on a "judgment of appearances" regarding the bearer of the self-ascribed state, the presence of character of the state, or its intentional content.

It might be useful to consider an example. "I"-ascriptions that are immune to error through misidentification such as "I see a Schipperke" (made in the normal way) may be corrected on the grounds that I have mistaken a Spitz for a Schipperke. But they are protected from my having misidentified the subject. It makes no sense to say "someone sees a Schipperke, but is it I?" Nonetheless, even as I am protected from recognitional or identificatory errors because no recognitional or identificatory capacity comes into play (in the case of avowals), I still succeed in saying something *about* myself. In particular, in ascribing a property or state to myself—argues Bar-On—I can be said to tell or to know that it is I who is in the state or who exemplifies the property (of seeing a Schipperke). It is the absence of a recognitional judgment that "someone is me" that makes certain "I"-ascriptions immune to error through misidentification. Similarly, the absence of recognitional judgment makes certain "I"-ascriptions immune to error through misascriptions. I can succeed in ascribing a state or property to myself, and when I do I can be said to tell or to know that I am in that state or exemplifying that property. I may be wrong: my avowals can be subject to *expressive errors.* I can, for instance, avow being bored when I am not. But with expressive failures, I express a product (a semantically articulate representation that conventionally signifies that I am bored) which is inappropriate, or false, of the present condition to which I (attempt to) give vent. Bar-On denies that it makes sense for me to ask (in cases of avowals) "I'm feeling something about my present circumstances, but is it boredom?" or "I'm feeling bored but is it upon reflection of my present circumstances or is it about something else?"

Speaking My Mind is impressive in its range, the detail of its argument, the clarity of writing, and especially in its cogent overviews of various related topics, including discussions and criticism of Anscombe, Shoemaker, and Evans on the reference of "I"; externalism about content; and "transparency-to-the-world" and "no achievement" views of self-knowledge. Bar-On's development of expressivism—used to support the epistemic security of avowals as well as their presumption of truth—includes criticism of "simple" expressivist views that fail to accommodate semantic continuity; a rational reconstruction of how noncontentful natural expressions form a continuum with expressions whose products are semantically articulate; as well as a careful attempt to distance her fully cognitive (though nonepistemic) account from the traditional noncognitivism that accompanies ethical expressivism. In the penultimate chapter, Bar-On defends her neo-expressivist account as one that will allow avowals to issue in genuine self-knowledge and in the final chapter she flirts with, but seems to me ultimately to offer reasons for rejecting, the idea that her account is consistent with functionalism or type-identity theories; indeed, her expressivist treatment presents these views with rather important challenges. And, because I shall end on a critical note, it is worth emphasizing here just how valuable this book should be for those working on self-knowledge (especially as the topic is situated within the contemporary philosophical literature), as well as for those interested in expressivism in its various domains.

One persistent qualm I had in reading the book was with Bar-On's description of her work as the development of an "account" of the epistemic security of avowals (one that fares better than others for various reasons, e.g., as it applies to both phenomenal as well as intentional avowals). This way of putting it makes it seem as if avowals form a recognizable kind, which is independent of the description of their epistemic security she offers. They do not. Utterances of the form "I [*mental verb*] *P*" count as avowals only on certain occasions of the utterance and these occasions are not specified independently of the description and explanation Bar-On develops of the ways in which they are immune to error. For instance, if we can imagine an occasion in which it makes sense for me to say that I feel something, but wonder if what I feel is boredom, such a case turns out to be a (perfectly admissible) act of self-interpretation—though perhaps protected by some kind of immunity, not the kind that attends avowals, and thus not an avowal. So Bar-On's "account" of the epistemic immunity of avowals, like an expensive bracelet, comes with its own safety-clasp: the view itself helps to fix the shape of the philosophical concept of avowal and thus is immune from criticism that it fails to explain various features of avowals, since features it cannot explain turn out not to fall within the extension of the concept. I thus think it is more appropriate to see the project as one that helps define the philosophical concept of "avowal," which we can then assess on those merits Bar-On claims for it.

This brings me to a more serious issue: the merits Bar-On claims for her treatment of avowals. It is not only important for her argument that her account of avowals' security includes all avowals—intentional as well as phenomenal—but Bar-On also insists that her account be consistent with a "full blooded realism" about mental states and with the possibility that avowals represent privileged self-knowledge.

In order to see the difficulty with these aims, consider what we are to say about the utterance "I understand." If this utterance is an avowal, then what can the avower be credited with knowledge *about* when she says she understands? That she understands? But surely that depends on what she is claiming to understand. Often, whether someone understands something (say, the meaning of a word) is a matter of what she goes on to do with it (viz., in certain circumstances, whether or not she uses the word correctly). In saying she understands, then, is she making a conjecture about what she will go on to do? If so, then this is not something she can avow. For no one can know the future: any number of things might prevent her from using the word correctly.

Bar-On recognizes a related difficulty for self-ascriptions such as "I am seeing a tree," which (on one understanding) is not an avowable state (e.g., on that understanding in which the truth of the claim depends on there being a real tree as opposed to one's merely having the impression of seeing one). But there are nonavowable uses of *all* intentional mental predicates: there are certain understandings of "understands," "intends," "hopes," "wishes," "believes," "wants," and so forth, in which what counts as satisfying these predicates is, in the appropriate circumstances, something about which the subject's pronouncements are not authoritative, since it is something about which she can merely conjecture (such as what she will go on to do, given the appropriate circumstances). (There is arguably a parallel story to be told about "phenomenal states," but I will not pursue this here.)

These facts—made familiar by Wittgenstein—indicate a problem for a certain construal of what is required of a "robust realism" for mental states. When it comes to understanding our practice of ascribing mental predicates, the Cartesian's favored epistemic route via introspection is not the only path to encounter obstacles. Bar-On may be commended (following Wittgenstein and Ryle) for drawing our attention away from the picture in which the self-ascriber is thought to be taking a considered look inside her head and reporting on what she sees (on one understanding of what that might mean) and toward a picture in which she is rather expressing herself instead (whether by "natural behavior" or by producing a semantically articulate, conventional representation). But if we say that in avowing, she is expressing her (antecedently existing) "state of mind," we are left with the same predicament that confronts the Cartesian. What state is this? The state of understanding? The expressivist picture is no better than the "inner theatre" model if both construe all uses of "I understand" (qua product) as the attribution of a kind of mental state or property. It is this construal that makes a mystery out of one important (if not paramount) use of the expression "understands": a use in which the condition that counts for the truth of that ascription (or for accepting it, justifying it, explaining what one means by it, etc.) is what the self-ascriber goes on to do—sometimes long after she says she understands. Again, that this future condition will obtain is *not* something she is in a position to avow.

What are we to make, then, of the immunity from error of the use of the expression "I understand" when it is a genuine avowal (thus protected from epistemic criticism) and *not* used as a conjecture about subsequent events? And how is this to be reconciled with its use when it is subsequent events that are decisive? This is the problem that exercised Wittgenstein from the early 1930s:

> One of the chief difficulties we have with the notion of a general idea or with understanding a word is that we want it to be something present at some definite time, say, when the word is understood, and the idea we have is supposed to have consequences and to act as time goes on.[8]

Avowals' immunity from epistemic error is a mystery not just on the "reportive," "descriptive," or "epistemic basis"-view of our relation to our own minds. It is a mystery on any view that construes (say) understanding (on one important use of "understanding"—say, understanding a word) as something "static, a disposition in the mind . . . ," or "as if the use of a word is like pulling a thread from a bobbin: it is all there and needs only to be unwound."[9] But this seems perilously close to the "robustly realist" view that Bar-On wants her account of avowals to accommodate.

In the final chapter Bar-On touches on these problems in the context of considering some of the ontological ramifications of the view. Here, she (reluctantly) comes close to accepting that her neo-expressivism is incompatible with a functionalist/materialist account of the nature of mental states. "[T]he special status of avowals and first-person privilege [developed in the book] are . . . not apt to be visible from the materialist point of view."[10] She tentatively suggests that we may avoid some of

8. Ludwig Wittgenstein, *Wittgenstein's Lectures, Cambridge, 1932–1935,* Alice Ambrose, ed. (Oxford: Blackwell, 1982), 85.
9. Ibid., 83.
10. Bar-On, *Speaking My Mind,* 426.

these problems if we consider mental predicates to name conditions of subjects (rather than states within them).

> What we take the subject to express . . . are *conditions the subjects are in, not states that are in the subjects.* It is these conditions that are perceived [sic] by the seasoned users of mental terms in the learning situation. And it is these conditions to which their use of the relevant terms is keyed. The ordinary concept of a mental state, then, we might say, is the concept of a subject's condition that is expressible in various characteristic ways, rather than the concept of a state inside the subject that has such-and-such typical (causes and) effects, or the concept of a brute disposition to display certain kinds of behaviour.[11]

I do not see, however, how this helps, since her "robust realism" seems to commit her to something like functionalism nonetheless. In defending neo-expressivism against the charge of "behaviorist irrealism" she attempts to maintain the logical independence of the state expressed and the behavior expressive of it. But she does not address the nature of the relation between the state expressed and subsequent (what she would agree to be nonexpressive) behavior. It seems her realism commits her to the view that, in using the predicate "understands," we (or the subject) attribute(s) to the subject the state or condition of understanding of which her subsequent behavior is a causal, contingent manifestation. If so, we again confront the difficulty that on those many uses (say) of "understands" in which what counts for truth is subsequent behavior, the subject's pronouncements do not enjoy distinctive security and cannot be considered to represent self-knowledge.

There is no space to pursue the point, but here is an idea. Wittgenstein's investigation of avowals such as "I understand" or "Now I can go on" is the study of a phenomenon that helps give rise to the idea—mistaken, in his view—that understanding is a "bobbin"-like state whose intrinsic nature is to be discovered: something static, a disposition in

11. Ibid., 424.

the mind, whose thread (i.e., its propositional content and all its applications) is all there and only needs to be unwound. (Changing "state within a person" to "a condition of the person" does not change the problem deeply rooted in this picture.) People who announce that they understand do *not* have a priori knowledge about what they will go on to do, though they are protected from epistemic criticism when they express their feelings or emotions (as they are when they describe their mental imagery).[12] The lack of puzzlement (understood in one way), a feeling of relief, a mental image—perhaps of a picture, a map, or a formula—are all genuine contenders for the description "mental states" (as opposed to ersatz "bobbin states"). They are also, importantly, the considerations one might point to in order to justify or explain some uses of "I understand."[13] But this is not so for a large number of cases. And, for this large number of cases, it seems to me to make little difference whether *if* there is this kind of mental "accompaniment," it is expressed with spontaneity or puzzled over with interpretive motives: for the occurrence of impressions or images such as these cannot be what we agree to (in these large number of cases) when we agree with someone that she understands. On certain uses of "I understand" (e.g., the meaning of a word) if there are occurrences such as these they are better described as symptoms of understanding. On these uses it is simply a mistake to think that what is said is made true by the existence of a mental state.[14]

---

12. Whether it so much as makes sense to describe this as representing *knowledge* is a moot point which I will not take up here.

13. When I ask my students during lectures if they understand, and I expect them either to nod or shake their heads, I may be asking if they are feeling puzzled (have the impression of puzzlement), in which case they are protected from criticism about whether they understand in this sense.

14. When I give my students marks on the basis of their understanding, I am not concerned about how they feel; what counts as understanding in this sense is how they do in their exams (and the students' avowals are not authoritative as to whether they understand in this sense). The impression of clarity or lack of puzzlement (or, say, successfully explaining to others the gist of the lecture), however, might be taken as symptoms of understanding in the sense I am concerned about. There are also uses of "I understand" in which the utterance itself (and not the expressing or the reporting of some feeling or impression) is the only consideration that matters. On these occasions "I understand"

If this is on the right track, then perhaps it points to a reason why some have wished to make a distinction between expressing a state of mind and making a self- or other-ascription. When I explain the use of a word to someone who has not heard it before and she says "I understand," perhaps she is precisely not to be construed as using "understand" in the same way as (as semantically continuous with) what I do when, after testing her, I say that she understands. (Which is why we might say, with Ryle, that here she is speaking sincerely or truthfully, but demur from describing her as speaking the truth.) The "criteria" we are using, as Wittgenstein says, may be in this case completely different—though, on good days, they coincide (the "condition" she avows is a symptom of, and possibly often accompanies, understanding in the sense I mean). Because if she is not conjecturing about what she will (in the right circumstances) go on to do, and she is, in avowing "I understand," speaking from or expressing a genuine mental state or condition, then that is not the condition I speak of when I say she understands.

---

works like "I promise" insofar as there is nothing *else* we can point to (except the utterance itself, as said on the particular occasion), that would give us a reason, justify, or explain our acceptance of it as legitimate or sincere (e.g., when a friend, in accepting your apology, says "I understand"). This kind of case really does seem to be one in which (as Crispin Wright says) avowals enter primitively into the language game.

CHAPTER SIXTEEN

Conceptual Amorphousness, Reasons, and Causes

1. In their volume of essays on agency and action, John Hyman and Helen Steward explain in their editorial preface that philosophy of action has been dominated, throughout its recent history, by positivism and its critics. The logical empiricists such as Carnap, Neurath, and Hempel rejected the view, accepted at the time, that there is an "impassable divide" in principle between natural sciences on the one hand, and those of the mind, society, or culture on the other: the latter thought to be subjects imbued with meaning, requiring "empathic insight," "introspection," and other devices for "understanding the sense of meaningful structures."[1] Philosophers of the 1940s and 1950s, in their turn, opposed the positivist conception and thus the idea, propounded by Mill, that "the human sciences are comparable to the exact natural sciences in their infancy."

As Hyman and Steward explain, the landscape changed again in the 1960s:

1. See Carl Hempel, "The Logical Analysis of Psychology," in *Readings in Philosophy of Psychology,* vol. 1, Ned Block, ed. (Cambridge, MA: Harvard University Press, 1980).

> The logic of sentences used to report actions, and the connections between the concepts of agent, action, event, and cause have been examined more scrupulously than they had been previously. And the orthodoxy is now a combination of ideas drawn from the positivist and the anti-positivist traditions. Some philosophers regard this as a plausible synthesis; others regard it as an unstable compound.[2]

Today's orthodoxy in the philosophy of action involves a combination of ideas drawn from both the positivist and antipositivist traditions; and the most common expression of the synthesis, due largely to Davidson, involves the claims that reason-explanation is a species of causal explanation and that reasons are causes.[3]

I am one who thinks this an "unstable compound"; that Davidson has, in effect, conflated different kinds of explanation and that this has led to the reemergence of Cartesianism in the philosophy of mind replete with insoluble problems. Arguing for this is complicated by the fact that "cause" is, as A. I. Melden said, a "snare" word in philosophy.[4] Elizabeth Anscombe famously remarked that a philosopher would do well to give up using "good" in moral philosophy and replace it with more fine-grained notions suited to her purposes; the same could be said about "cause."[5] The notions of reason and cause blur together at the boundaries even when their uses are regimented to respect a distinction between different styles of explanation (and again it is Anscombe

2. John Hyman and Helen Steward, eds., *Agency and Action, Royal Institute of Philosophy Supplement* 55 (Cambridge: Cambridge University Press, 2004), v.

3. Indeed, for Davidson it seems that reason-explanation is a species of causal explanation insofar as a (primary) reason for an action is its cause (as he goes on to develop the idea). The two claims are made in the first few pages of his "Actions, Reasons, and Causes," (1963), *Essays on Actions and Events* (Oxford: Clarendon Press, 1980), 3–20. Whether or not this thesis is compatible with his views on the anomalousness of the mental, causal relations, and causal explanation (which for him, is deductive-nomological) is an important question. I register my suspicion that there is a problem in "Why Reasons May Not Be Causes" (chapter 5).

4. A. I. Melden, *Free Action* (London: Routledge & Kegan Paul, 1961).

5. This way of putting it was suggested to me by Bas van Fraassen.

who has given us examples of this);[6] but, more importantly, both words serve in ordinary discourse to signal an ill-defined explanatory context. Thus, before philosophical regimentation, we speak of the reason why moss grows in the shade or why water freezes in subzero temperatures as we speak of AIDS causing the suppression of the immune system, or of gravity causing the apple to fall from the tree.

We speak these ways, but there may be good reasons for regimenting language, for philosophical purposes, so that cause, reason, and the explanations in which they figure are delineated in such a way so as to make more precise different kinds of explanations at work. (This would at least give the positivist and her opponent something to argue about.) This classification seems already to be in play when Steward and Hyman claim that the orthodoxy today involves combining the ideas drawn from both traditions.

Not everyone agrees with this regulation of language, however. Indeed, the debate between the causalists and anticausalists in the domain of action and reason-explanation is made even more complicated by the fact that certain philosophers (whom I shall call "compatibilists") impressed by the arguments of Ryle, Anscombe, Wittgenstein (and their kin), and who agree that confusion may result from mixing up several different paradigms of explanation, nonetheless decline to fight the battle over the introduction of causation into an account of reason-explanation. The compatibilist's charge against the anticausalist is that she allows the opposition to "hijack" a notion of cause (which buys into some or other aspects of the positivist picture) which should not be conceded.[7] Consider, for example, the following passage from McDowell:

> Reasons must be capable of explaining actions; and it can seem that if we are to find that intelligible, we must conceive practical reason as directing the action-generating efficacy of a collection of motivations that are prior to it, and, as far as rational explanation

6. G. E. M. Anscombe, *Intention* (Oxford: Basil Blackwell, 1957), §15.

7. This complaint has been voiced, for example, by Jim Conant.

> goes, simply given. This idea is clearly congenial to the internal conception. But it rests on what looks like a misconception of the way actions are explained by reasons, one that assimilates that to the way events are explained by mechanical forces such as the tension in a tightly wound spring. One can be suspicious of this assimilation without threat to the thesis that reason-giving explanations are causal. [A footnote refers us to Davidson's "Actions, Reasons, and Causes."] To respect the truism that reason-giving explanations work by revealing how actions are motivated, it is not necessary to picture motivations as antecedent quasi-mechanical sources of energy: in explaining an action, a reason-giving explanation can equally make the motivation of the action rationally intelligible, and there is no basis for insisting that this must be by way of representing that motivation as the upshot of reason's channelling a pre-rational motivational force in a certain direction.[8]

One would need to know more about how he conceives of causation to be in a position to evaluate McDowell's claim that reason-explanation may still be regarded as causal even if one were to give up the assimilation of causal explanation with the quasi mechanical. I shall return to this at the end.

Let us consider in more detail a well-known philosophical account of the concept of cause.[9] Some, following Davidson, have suggested that causation be construed as an external relation (one that can be captured by an extensionalist logic; one that obtains in the world anyway, as it were, independently of what concepts we use to identify the relata); some have suggested that these relata are events (which are themselves subject to different understandings); that this relation is discovered by experiment and not by logic or philosophical reflection;

8. John McDowell, "Might There Be External Reasons?," (1995), *Mind, Value, and Reality* (Cambridge, MA: Harvard University Press, 1998), 108.

9. For criticism against a broadly Humean or 'empiricist' account of cause, see my "Prolegomena to a Cartographical Investigation of Cause and Reason" in *Reasons and Causes: Causalism and Non-Causalism in the Philosophy of Action* , Giuseppina D'Oro, ed., History of Analytic Philosophy Series, (London: Palgrave MacMillan, forthcoming).

that these relations instantiate laws; that the cause-event necessitates its effect; and so on. Some philosophers reject some or other aspects of this, by and large "Humean" picture: by construing causal relations merely as counterfactual-supporting; by allowing that facts and particulars (other than events, on one understanding) enter into causal relations; by conceding that causes need not necessitate their effects; or by accepting that singular causal relations need not instantiate laws, and so on. Nonetheless, according to Peter Hacker:

> [t]he dominant philosophical account of causation conceives of the causal relation as non-logical (external), inductive (hence requiring the possibility of independent identification of the relata), and nomic (instantiating a general law).[10]

Ryle accepts a certain picture suggested by a use of "cause" in which causal laws are established by experiment, and the relata that enter into causal relations are independently observable and describable. This he contrasts with a phenomenon that is partly constitutive of another, "somewhat as the headside of a penny is indeed part of what makes it the penny it is, and yet is obviously not a separately existing agency that causes the penny to be the penny that it is."[11] For Wittgenstein, a crucial contrast is between what is discovered on the basis of experiment vs. "the rules we lay down."

The goal of this chapter is to explore these contrasts in order—eventually—to cast light on the dispute between the causalists and noncausalists in the domain of reason-explanation.[12] In section 2 I

10. Peter Hacker, *Wittgenstein: Meaning and Mind* (Oxford: Blackwell, 1990), 135.

11. Gilbert Ryle, "A Rational Animal," (1962), in *Collected Papers,* vol. 2 (London: Routledge, 2009), 421. See "Reasons as Non-Causal, Context-Placing Explanations" and "Pain, Polio, and Pride: Some Reflections on 'Becausal' Explanations" (chapters 7 and 8) for my own suggestions about how to understand this distinction.

12. One matter for discussion elsewhere is the extent to which "compatibilists" are committed to this contrast when they allow that Davidson's arguments may still be honored. Jennifer Hornsby maintains that understanding the agent's role—as cause of what her actions cause—allows us preserve the idea that reason-explanation is a species of causal explanation. Jennifer Hornsby, "Agency and Actions," *Agency and Action,* Hyman and

shall introduce a view about the way language functions, shared by Wittgenstein and Ryle, which brings to light what I call the "amorphousness" of concepts that are discerned from natural language expressions. In section 3 I shall illustrate how this view can be applied to the notions of 'love' and 'understanding' and, in section 4, how it can be applied to the concept of reasons for acting. With this view of language in place I shall argue that if the causation is construed as an external, inductive relation, then reasons can only in relatively rare cases—and with a certain stretch—be glossed as "causes." But this is enough to show that the explanatory power of reason-explanation is not causal. Because part of my aim is to bring Wittgenstein and Ryle back into prominence (as well as a certain way of engaging in philosophical enquiry) my discussion is peppered liberally with citations from their lesser-known works (the former's *Cambridge Lectures* and the latter's *Collected Papers*).

Steward, eds., 1–23. In saying this, as I note below, she departs in crucial ways from the way Davidson develops his argument in "Actions, Reasons, and Causes." This may be (at least in part) because she (with Davidson, but unlike me) construes "she did such-and-such because she desired . . ." as an example of "causal statements . . . that hardly need defense; they are statements of a kind that we commonly recognize to be true" (Hornsby, "Agency and Actions," 8). My approach, by contrast, tends to point out the differences in "becausal" explanations so that causal ones are distinguished, for example, from constitutive, meaning, normative, or grammatical, ones (in spite of Davidson's arguments to the contrary). Both Hornsby and I argue against the "standard" story of action explanation (roughly characterized by the slogan that beliefs and desires cause actions). The problem for Hornsby is the "events-based character of the standard story which gives rise to the idea that action-explanations record truths about causally-related particulars" (ibid., 9). My tack has been to point out that there is little support in our reason-attributing practices that should incline us identify reasons with anything hidden or inner (e.g., physical properties of the agent) that are candidate causes of her action (see section 3). Since the standard story that reasons are such causally efficacious items is one I think that Davidson's arguments in "Actions, Reasons, and Causes" commit him to, it perhaps should be made explicit that the arguments in this chapter against the thesis that reasons are causes operate with a different (and I hope uncontroversial because neutral) understanding of "reason." The main aims of this chapter are to consider how Wittgenstein and Ryle might help us to resist the assimilation of reasons (in this sense) with causes and to point out the difference in vocabulary that makes some of the compatibilists' claims seem so foreign to some of mine.

2. Wittgenstein famously noted a family of structures, more or less related, that can be discerned in the use of natural language expressions, as opposed to the existence of "some one thing in common which makes us use the same word for all." As a result of the kind of philosophical investigation he recommends "we see a complicated network of similarities overlapping and criss-crossing: sometimes overall similarities, sometimes similarities of detail."[13] Ryle, for his part, suggested the philosopher's job is to trace the "elasticities of significance" and "inflections of meaning" in expressions, which, he tells us, affect the inferences that one is allowed to make from, say, one occasion of utterance to another (influenced by the kind of thing talked about, our purposes or needs, and other aspects of the circumstances). The *Philosophical Investigations* invites the reader to perform a variety of exercises to bring this point home; Ryle gives many examples of this phenomenon in his *Collected Papers* as well as in *The Concept of Mind;* and more recently, Charles Travis's work provides a number of illustrations on the theme.[14] I shall illustrate this phenomenon, however, with an example of my own.

The late, great "Mollie Munchkin," my feline companion for seventeen and one-half years, was a solid red Persian, descended from a line of Grand Champions on her father's side. Her "sister," Ivy, was a gorgeous Blue "Exotic": a cross between British Blue and Persian. One can tell from looking at photographs that Mollie's coat was the same color as the pale oak floor of our century-old Parisian flat. Mollie's coat was red. It was the same color as the oak floor. But the oak is not red: it is light-to-medium brown.

The word "red" undergoes a particular inflection of meaning when it is used to describe cat fur (or human hair) for example, as opposed to wood (or sand, clay, etc.) and many other substances and inanimate objects. This particular inflection foils a normally acceptable inference: in this case, one is not permitted to infer that the oak floor is red from

---

13. Ludwig Wittgenstein, *Philosophical Investigations,* G. E. M. Anscombe, trans. (Oxford: Blackwell, 1953), §§65, 66.

14. Charles Travis, *Unshadowed Thought: Representation in Thought and Language* (Cambridge, MA: Harvard University Press, 2000).

the fact that the cat is red, even on the assumption that the cat and the oak are the same color.

The logical inflections of natural language expressions that come to light when used in different circumstances point to a phenomenon that I shall call "conceptual amorphousness," which if pervasive, should have consequences for philosophical explorations or investigations of the abstract phenomena that are illuminated by these concepts. Before discussing this in more detail, let us first answer a number of objections to the claim that my cat example shows any interesting facts about language.

First, one may be tempted to suppose that the example simply equivocates on the word "red" since it is used ambiguously. No philosophically interesting conclusions about meaning can be derived from the use of an example that exploits the kind of ambiguity that logicians ignore, for these sorts of wrinkles are mere artifacts of natural language which should be ironed out of an idealized one.

We shall revisit the thought that an idealized language should "iron out" this kind of ambiguity in a moment, but note, first, that this example does not exploit a word that is robustly ambiguous, at least not in the way that words like "bank" or *"restauration"* are. The definitions of words that are ambiguous in the latter sense (Ryle calls them "pun-words") are completely and arbitrarily different from one another. Words with different inflections of meaning are, by contrast, unlike "pun-words," they are intimately connected with one another: "they are, in one way or another, inflections of the same root."[15] In our example, "red" is used consistently as a color word. With this aspect of the meaning fixed, the examples remind us that what it is for cat or human hair to be red is something other than what it is for many other objects to be red. (And what it is for human skin to be red (e.g., inflamed) is different again.)

Second, one may be inclined to dismiss the example by conceding that color words in particular are applied differently depending on the

15. Gilbert Ryle, "Philosophical Arguments," (1945), in *Collected Papers,* vol. 2 (London: Routledge, 2009), 215.

objects they describe, much as the names for groups of wildlife change to suit the kinds of animals or fish that are assembled. That we may replace "school" with "pod" to talk of groups of dolphins instead of fish or whales is a relatively uninteresting feature of English that one would not suppose to carry over to other areas of discourse, let alone to an idealized language.

In response, it should be conceded that the example exploits this fact about color words and their objects, but the "systematic ambiguity" or "circumstance-dependence" of the expression reappears even if we put this fact aside. Suppose that everyone present at a cat show is asked to remove red objects from sight since a runaway bull from a neighboring event has entered the arena. Correctly obeying this request, presumably, would not involve hiding Mollie as opposed to Ivy, though it might if the former were wearing a special (e.g., Christmas) outfit. This suggests that the uses of "red" in question here do not simply track kinds of objects (certain cats), in special circumstances (cat shows), for even when cats feature among the items described, and even if cat shows figure in the background, we can think of particular circumstances in which it is false to say of those cats, in those circumstances, that they are red.

Third, one may be inclined to dismiss the example by claiming that the use of "red" to describe Mollie or her ancestors is metaphorical, for they were not really red.

In the context of a cat show in normal conditions—for example, when a bull is not threatening to run amok but rather, say, while the judge is examining her or her forefathers—the remark would be false. Her father and grandfather were Grand Champions, and they would not have achieved this distinction unless they were really red.

One cannot explain away the feature of language my example brings into focus simply by appealing to what is real, since what we call *real* (along with what we call *red, solid,* and *is the same color as*) depends on various factors. When it comes to cat colors, the rules for what counts as red (as well as for blue) may be different, depending on the occasion, from what counts as red or blue for other objects or even the same objects in other circumstances.

I have chosen this example because it is clear that the inferential anomalies brought to light can be explained by pointing to a linguistic convention about what counts as being, say, red or blue in cat fanciers' associations. A long time ago it was laid down as a rule, or a convention was established, that cats the color of Mollie count as solid red. We should agree that this example uncovers no interesting physical or metaphysical facts about color properties.

Ryle thought that this sort of elasticity of signification characterizes the use not of a *few* but of *most* or *all* expressions. For almost every word or phrase we use contributes to what we say in a way such that had it been replaced by a word or phrase with a different inflection, it would have had different implication threads:

> It would have been a different statement, different in having different implications, in requiring different tests for truth or falsehood, in being compatible and incompatible with different affiliated statements, in being evidence for or against different corollaries, and so on.[16]

This is true not only for words but for complex expressions and grammatical constructions as well.[17] The capacity for words, phrases, sentences, and grammatical constructions to have the power to express an indefinite range of differing logical types, and therefore, to enjoy different logical powers is systematic; their capacity "to acquire new inflections of logical forces is one of the chief factors making original thought possible."[18]

This has implications for the study of concepts. The idea that concepts exist in "isolated splendour" and that they are the objects at which our natural language expressions merely gesture gets things back to front on the view I am promoting. On this view, concepts exist, as it were, only as abstractible features which can be discerned from the functioning or working of the expressions from which they take form. If these expressions function to express an indefinite variety of

---

16. Gilbert Ryle, "Abstractions," (1962), in *Collected Papers,* vol. 2. (London: Routledge, 2009), 456.
17. Ryle, "Philosophical Arguments," 215.
18. Ibid., 216.

ideas, then the concepts which are thereby discerned will have a certain amorphousness, as I see it, like a soap bubble whose contours change depending on the atmosphere, rather than a pure crystalline structure that preserves its shapes across contexts and purposes.

In drawing attention to this amorphousness I am, in effect, questioning the assumption (along with Ryle and the later Wittgenstein) that single meanings governing their logical behavior can be attached to single expressions across contexts. Indeed, even within a context, broadly defined, there may be several, perhaps indefinitely many, possible meanings or implication threads of these expressions, depending upon the particular ends or interests of those wielding them. It is doubtful, then, that the logic of these statements can adequately be represented by imagining they have one function. As Ryle puts it, both the nonformal expressions of everyday discourse as well as those of technical discourse have their own unscheduled logical powers. The only way to uncover these is to "pull" at the implication threads of the expressions as their uses are explored from one set of circumstances to another. This is the nature of a conceptual investigation.

Both Ryle and Wittgenstein thought that the failure to appreciate what I am calling "conceptual amorphousness"—which is a function of the capacity of natural words, phrases, etc., to express an indefinite variety of ideas—is the main cause of philosophical perplexity, and that correspondingly, anomalies that lead to such puzzles can be dissolved by pointing to the diversity of conventions that underlie language. The way to investigate the nature of the particular abstract phenomena which interest us is to investigate the corresponding concepts. The way to explore these is to look at the assorted circumstances in which we deploy the terms or expressions from which the concepts are discerned. Failure to pay heed to these inflections may, according to Ryle, result in confounding the logical force of the expression from one occasion of utterance to another. (It is this confusion of force that Ryle calls a "type-error" or "category mistake.") Wittgenstein complained that we are continually gripped by various pictures that force themselves upon us as a result of focusing too narrowly on some uses of language to the exclusion of others. One perennial way of being misled is to assume that language functions in the main to report or describe states of affairs and that it does so by attributing a quality, property, or relation to an object, state, or event. The way to resist the philosophical problems that result from being thus gripped is to remind ourselves of the wider set of circumstances in which the relevant expressions are correctly and appropriately used. Again, the realization that these expressions have a multitude of uses, including new ones, should convince us of the futility of trying to investigate the nature of the particular abstract phenomenon without studying the vast number of functions performed by the expressions from which the concepts are abstracted as discernable features.

A point about regimentation of language is in order here. Philosophers—Wittgenstein and Ryle included—propose classifications or regularizations of language all the time. The idea is not that we cannot or should not fix the contours of a concept for certain ends. For example, we might wish to mark only those affinities between expressions that we need for particular purposes, for example to hold them up as a point of comparison, or to suggest that it is these features which are paradigmatic and should have a privileged seat, as it were, in the center of the conceptual manifold. We are, after all, fixing the contours of causation in treating it in accordance with the dominant picture in order to bring out the difference between kinds of explanation. We shall also be exploring key uses of the expression "reason for acting" and its cognates, which will help demarcate the concept and thus the phenomenon of acting for such and such reasons. Doing this, however, is not to make the mistake that is one of the targets of this essay, which is to imagine that concepts have a crystalline structure which, existing independently and logically prior to them, determines the correct use of their proper expressions.

3. Just as there are different inflections of significance of "red," so too are there different inflections of "reason," "intention," and "understanding." There are different criteria—different kinds of grounds—for saying that somebody acted for such-and-such reasons, or with such an intention, or with understanding. As I understand it, these grounds are part of the landscape of the concept. Closest to the center the grounds may overlap. Closer to the edges they may not. Sometimes when the backings and support for the claims in which these concepts are "expressed" are different we have to make a (reasoned) decision about which are appropriate for the occasion—in effect, we need to locate the place on the landscape that our use of an expression, on a particular occasion, occupies. Failure to appreciate that it may occupy a place in which its grounds and backing are different from that of another may lead us (mistakenly) to assimilate the logical force of an expression when it is used in one way with its force when it used in another. A philosopher who does not accept that there may be no one thing that ties these uses of expressions together may be tempted to construct a theory that tries to accommodate the various uses while at the same time running the risk that her account will fail to do full justice to any.

As an example let us suppose, adapting an example Wittgenstein uses in his lectures,[19] that Michael has all the normal feelings toward Paul that would incline us to say that Michael loves Paul, but, when given the opportunity later on, he does nothing to save Paul's life. In such a case we might naturally say that Michael did not love Paul after all, but in saying this, Wittgenstein suggests that we are not (necessarily) contradicting what we said earlier. Nor is it the case that we are prepared to withdraw our description of Michael as loving Paul because of an empirical regularity between loving and saving lives which Michael has failed to instantiate. Rather, we are reasoning about what we will *call* or *count as* love in the circumstances. We are required to do this when the backing and support for a claim about love would lead us in the one case to apply the concept and in the other to withhold it. This is not a mere feature about language, however. Tracing the inflections of significance or the logical threads of the concept of love *is* to investigate its nature.

If one is seduced by a picture that says there must be some underlying thing (a property, a set of necessary and sufficient conditions) that ties the various uses of the relevant expressions together (from which the concept of love is abstracted as a discernable feature), one will have

19. Ludwig Wittgenstein, *Wittgenstein's Lectures, Cambridge, 1932–1935,* Alice Ambrose, ed. (Oxford: Blackwell, 1982), 90–91.

the problem of trying to reconcile these uses and thus in understanding the concept. On the one hand there are "static" tests for love, as there are for understanding, and intending which (say) may have something to do with the way the person is feeling or thinking. On the other hand there are tests that extend over time for these phenomena. For many of our ordinary mental concepts—those, for example, of special interest to philosophers of mind and to action theorists, there is also a third criterion—namely, what the person, herself, spontaneously utters or avows as well as what she says in reflective self-interpretation. These criteria may overlap in certain interesting ways, but they need not. When they do not, we should be prepared to treat these as separate or different considerations. In the cases where we have to decide between competing criteria, we let the circumstances and our purposes or interests influence our decision.

Not accepting that there are different criteria for love, and insisting that there must some one thing which ties the uses of the word together, we may be tempted to think of love as a disposition. Consider how Wittgenstein describes this mistake:

> A disposition is thought of as something always there from which behaviour follows. It is analogous to the structure of a machine and its behaviour. There are three different statements which seem to give the meaning of "*A* loves *B*": (1) a nondispositional statement about a conscious state, i.e., feelings, (2) a statement that under certain conditions *A* will behave in such-and-such a way, (3) a dispositional statement that if some process is going on in his mind it will have the consequence that he behaves in such-and-such a way.[. . .] We seem to have distinguished here three meanings for "*A* loves *B*," but this is not the case. (1), to the effect that *A* loves *B* when he has certain feelings, and (2), that he loves him when he behaves in such-and-such a way, both give the meanings of the word "love." But the dispositional statement (3), referring to a mechanism, is not genuine. It gives no new meaning. Dispositional statements are always at bottom statements about a mechanism, and have the grammar of statements about a mechanism. Language uses the analogy of a machine, which constantly

> misleads us. In an enormous number of cases our words have the form of dispositional statements referring to a mechanism whether there is a mechanism or not. In the example about love, nobody has the slightest idea what sort of mechanism is being referred to. The dispositional statement does not tell us anything about the nature of love; it is only a way we describe it. Of the three meanings the dispositional one is the only one that is not genuine. It is actually a statement about the grammar of the word "love."[20]

In considering love a disposition in this sense—a process or state of mind which has the consequence that one behaves in such-and-such a way—we are conflating two different things: a process in our minds or brains, whose causes and effects can be studied by psychological methods and "rules which we lay down."[21] A similar temptation arises with the notion of understanding a word or idea.

> One of the chief difficulties we have with the notion of a general idea or with understanding a word is that we want it to be something present at some definite time, say, when the word is understood, and the idea we have is supposed to have consequences and to act as time goes on. For example, the idea of a plant is supposed to enable me to identify something as a plant, bring a plant when ordered to, define "plant," etc.; and these phenomena are taken to agree or disagree with the idea. If by "general idea" we mean the *cause* of agreement and disagreement, there would be no difficulty, for then the idea would be an existent thing like an acid to which there is a reaction of some sort. But we do not want the relation between the idea and a phenomenon which agrees or disagrees with it to be a mere causal one. The agreement we want is not experiential at all. It is not a question of experience whether a thing will agree with our general idea, as it is with a mechanism about which we cannot predict with certainty. If we take the idea to be a natu-

20. Wittgenstein, *Cambridge Lectures, 1932–1935,* 90–91.
21. Ibid., 84–85.

> ral phenomenon which can do such things as enable us to apply a general word or give a definition of it, our investigation of it is psychological. We are in the realm of hypotheses, about effects and causes, and not in the realm of the "must." But we are wrong to take the investigation of the general idea to be an investigation of the causes and effects of a natural phenomenon. We are mixing up two different things, a process which happens in our minds or brains, whose causes and effects can be studied by psychological methods as in other sciences, and certain rules which we lay down.[22]

When Wittgenstein talks here about "rules we lay down" and distinguishes an investigation of these from that of the causes and effects of a natural phenomenon he means to be drawing our attention to the feature of language my cat example was intended to demonstrate. "Understanding," like "red" and "love," undergoes inflections of meaning, as different kinds of considerations or criteria explain or justify our claims in which these concepts are wielded. These inflections may suggest that understanding is both a static state and something that has consequences in the future. Forgetting that these considerations may point to two different uses of the word we might be inclined to assimilate the uses and say, for example, that it is my understanding of a word that enables me to use it correctly (or in accordance with an idea that I have grasped). This suggests that understanding is a disposition—understood here as kind of mechanism—an inner state with causal effects. But in accepting this suggestion, Wittgenstein indicates, we are refusing to see that these may simply be two different uses of the word which we have mistakenly assimilated because the grammar forces the picture of a dispositional state (qua mechanism) upon us. Ryle, in his turn, argued against this very assimilation in his attack on the "intellectualist legend."[23] In construing understanding as a dispositional state or mechanism, we are, according to Wittgenstein, conflat-

22. Ibid., 84–85.

23. It is worth a word of warning that although Ryle also talks about the dispositional aspect of mental concepts, he is interested in the way sentences containing mental verbs can be "unpacked" by "hypothetical" and "semi-hypothetical" sentences and specifi-

ing two things: a rule we lay down about what we will call or count as understanding and a natural phenomenon. The mistake made by one who treats understanding as a dispositional state or mechanism ripe for empirical investigation is on a par with that made by one who would construe my cat example as showing the need for an empirical investigation of color properties in order to tell us more about how they change across contexts.

Consider how we might elaborate some different uses of "understanding." If I ask my students if they understand the concepts I have been teaching them as *part of a revision lesson* then it would be natural to construe me as asking them to predict or form a hypothesis about their future behavior in the exam. If I ask my students (in the middle of a long, complicated lecture at the beginning of the term) if they understand what I am saying I *may* only mean to ask them if they are able to follow my line of reasoning then and there: or indeed, if they are confused. In such a case, they are not inconsistent if they say that they do understand, but are unable to repeat the ideas back to me or if they do poorly in a subsequent exam. If I ask a dying friend if she understands my reasons for not having been in touch, what *may* matter is just that she be willing to say she understands (and no future test is relevant).[24] A similar variety of criteria is available for what we count as reading, as Wittgenstein discusses at length in the *Investigations* §157–178.

4. SO TOO is a similar diversity of considerations available for what we count as acting for reasons. Consider what Wittgenstein says about reasons in his lectures:

> To do a thing for a certain reason may mean several things. When a person gives as his reason for entering a room that there is a lecture,

cally denies (agreeing here with Wittgenstein) that these sentences are to be construed as picking out a causal mechanism.

24. Compare: when I ask my husband if he loves me what (usually) matters is that he simply be willing—indeed happy—to say so (and to say so in a certain tone of voice). In another context—say, if we were contemplating whether to stay together—reflection about his actions and reactions into the future would be relevant.

> how does one know that is his reason? The reason may be nothing more than just the one he gives when asked. Again, a reason may be the way one arrives at a conclusion, e.g., when one multiplies 13 x 25. It is a calculation and is the justification for the result 325. The reason for fixing a date might consist in a man's going through a game of checking his diary and finding a free time. The reason here might be said to be included in the act he performs. A cause could not be included in this sense.[25]

To elaborate on this, it seems that we have a decision to make, in any particular circumstance, about what we will count as *the reason for which* so-and-so acts. If what I care about is the "narrative" or story the person uses to make sense of herself then I will be interested in what she tells me. Often what she tells me will also make her action intelligible. We have seen in other chapters why this should be so.[26] Her ability to say why she acts is conditioned or constrained by training in a shared discourse and a way of life in which what she does and what she says are revealed to be intelligible. But there may be occasions in which the answer she offers fails to meet such demands. There are controls, as Anscombe says, on the truthfulness of someone's proffered reasons.[27] If what she says does not make her action intelligible, and if being able to make sense of it matters more than her own "narrative," then I may reject what she tells me in favor of the criterion that allows me, an onlooker, to make sense given the circumstances and the behavior I observe. (Notice that this criterion allows me to use my own vocabulary to explain why she acted and not necessarily one that she shares.) If, however, we want to know what procedure she followed to reach, say, a decision to act then we may again be dependent on what she says if this procedure was performed, say, silently in her head. Or not, if there are other criteria we can rely on for the higher-order activity, as there would be, say, if she had recorded her deliberations in a diary.

---

25. Wittgenstein, *Cambridge Lectures 1932–1935,* 5.
26. See the earlier chapters in part 4 for more discussion.
27. Anscombe, *Intention,* §25.

I have considered three (rather general) criteria for saying—or grounds for claiming—that an agent acted for such-and-such a reason. On one understanding, a reason can be read off the circumstances in which the action takes or took place.[28] Here the reason gives us a way of making sense of the performance. This is an instrument, as it were, used at the level of discourse that belongs, as Ryle suggests, to the theorist or referee. This instrument, used at this level of discourse, may or may not have been used by the agent when she acted.

On another understanding, "the reason may be nothing more than just the one he gives when asked." Here, what a person says may or may not involve a memory report about a particular procedure followed in the course of her acting. "What is said" may simply count as a *distinct* criterion, in appropriate circumstances, of what we mean when we ask for someone's reasons.

The third understanding—"the way one arrives at a conclusion"—does invite us to look for a certain procedure that is followed in the course of a performance. And, as with the other criteria, this one may or may not overlap with the others in the sense that different kinds of grounds may be given. If the procedure involves an appeal to reasons then this understanding overlaps with the first insofar as it employs reasons as a sense-making tool. It overlaps with the second insofar as this employment may be reported or described by the agent when she tells us her reasons.

Here, in this last example, the reason is appealed to as part of a procedure which culminates in action. Perhaps reasons may be causes on this (rather stretched) understanding in the sense that it is in virtue of the subject's accepting that such-and-such consideration favors a particular type of action that she is moved to act as she does. Here may be a sense in which reason-explanation can reveal how actions are motivated without, perhaps, as McDowell wishes, picturing motivations as antecedent quasi-mechanical sources of energy.

---

28. I describe this consideration in some detail in "Reason-Explanation and the Contents of the Mind" (chapter 6).

Let us for the sake of argument accept that one understanding of what it is to act for such-and-such reasons may involve some such story: this is the "calculation" criterion, perhaps here taking the form of a deliberation procedure or an explicit appeal to reason that accompanies or precedes the action. Whether the calculation is performed correctly and in that sense justifies the action is another matter, though bad calculations, like false beliefs, can also make certain puzzling actions intelligible. This is rather far removed from the standard picture of causation, but at least we can make out the existence of two independent events. Perhaps, then, if there is some sort of procedure which precedes an action that involves calculating, deliberating, or weighing reasons, we may follow Anscombe in calling these "mental causes" (which, as a class, may or may not include a justificatory element).

> A mental cause is what someone would describe if he were asked the specific question: what produced this action or thought or feeling on your part: what did you see or hear or feel, or what ideas or images cropped up in your mind, and led up to it? I have isolated this notion of a mental cause because there is such a thing as this question with this sort of answer, and because I want to distinguish it from the ordinary senses of "motive" and "intention," rather than because it is in itself of very great importance; for I believe that it is of very little. But it is important to have a clear idea of it, partly because *a* very natural conception of "motive" is that it is what *moves* (the very word suggests that)—glossed as "what *causes*" a man's actions etc. And "what causes" them is perhaps then thought of as an event that brings the effect about—though how it does—i.e. whether it should be thought of as a kind of pushing in another medium, or in some other way—is of course completely obscure.[29]

In coming to understand why someone acted as she did, a description of events such as these *may* be an important part of the story

29. Anscombe, *Intention,* §11.

(think of Camus's Meursault). But of course there may not have been any such events and even if there were in a large number of cases we should agree with Anscombe that such mental causes are of relatively little importance. In any case the domain in which reasons explain action extends far beyond the rare case in which action is preceded or accompanied by calculation or deliberation. For there may not have been any such procedure, and yet we may still give grounds for our claim that a person acted for such-and-such reasons and that in so acting she was manifesting her "full-blown" agency and autonomy. For in seeking her reasons for acting, it may be that what matters to us is what she says about her action; or it may be that what matters to us is that we are able to see the sense of it in the circumstances. What if we cannot make sense of her action and if in response to our questions she is unable to shed any light? In such a case, if there is any meaning left to "the reason by which she acts," its application needs to be shown. The danger is in thinking that the calculation model needs to be recast as the model into which all reason-explanation must fit.

To sense how the fatal temptation arises to view reasons as dispositions in Wittgenstein's sense—as a process going on in the mind that will have behavioral consequences—consider the following. If it is supposed that the concept of acting for such-and-such a reason picks out one single phenomenon, then the following moves will be appealing. First, what the agent says will be rejected as determining her reason for acting on the grounds that this is sometimes overridden by considerations that make more sense of what she does. Second, what makes sense of the behavior in the circumstances will be rejected as determining her reasons on the grounds—similar to those given by Davidson—that this may lead to an indeterminate result or, if the circumstances are too narrow, it may not rule out coincidence or accident. And third, the calculation or deliberation criterion will take pride of place since this includes both the sense-making feature (as that which would justify the action) and what the agent would say about her reasons for acting, had she been asked and had such a procedure or appeal occurred. But when faced with the fact that often a person acts for reasons without calculating or deliberating in advance of or during the action, one may be tempted to imagine these grounds *must* be available and therefore posit unconscious or tacit acts of calculation. One may thus be inclined to conceive rational action as issuing from some kind of dispositional state cum mechanism. Perhaps it is this picture that motivates the causalists. If so, it needs to be resisted.

I have (for the sake of argument) allowed that on one understanding of what it is to act for certain reasons, reasons and causes may converge. This is the calculation criterion: a subset, because it includes a "justificatory" element, of the class of events Anscombe calls "mental causes." But there are other understandings of what it is to act for certain reasons which do not require calculations or deliberations or even explicit thinkings-about-the-situation-ahead-of-time. These are other understandings in which reasons are not causes, and reason-explanation is not a form of causal explanation. It should not be supposed that in the cases where there has been no explicit calculation, deliberation, or appeals to reason preceding or during the action, that this must have gone on "implicitly" or "tacitly"—especially if these latter terms suggest the existence and operation of some inner mechanism.

One welcome consequence of viewing these as separate and possibly nonoverlapping understandings of acting for reasons is that it allows us to accommodate the idea that there is something special about human beings: for they, as language speakers, can satisfy all of the (above-mentioned) criteria for acting for reasons. (Although it should go without saying by now that satisfying all three cannot be thought necessary in order for a person to manifest her full-blooded agency on every occasion.) Another welcome consequence of doing so is that it accommodates the commonsensical conviction that nonlanguage users act for reasons insofar as they satisfy the criterion that we can make sense of what they do by observing them.

Accepting these as separate and possibly nonoverlapping criteria for acting for reasons also has important implications for the topic of self-knowledge. Crispin Wright has suggested that avowals enter primitively into our ordinary practice of cognitive/psychological explanation: that the default position is to accept what the avower says, pending another explanation that makes her actions more intelligible;

and he has suggested that this primitive feature of our everyday practices is sufficient to explain the avower's authority.[30] There is something misleading about this since, as I remarked above, it is no accident that avowals and intelligibility coincide: the avower is a participant in our sense-making practices and her ability to say why she acts is conditioned by training in a shared discourse and a way of life in which what she does and what she says are revealed to be intelligible. In this respect, avowals are not primitive; they depend on a common framework of what counts as intelligible. But there is also something correct about this way of putting it too: because what it *is* to have acted for certain reasons may be constituted, according to *one* of the criteria we are examining, by what the agent says.

Finally, accepting this may allow us to demystify the philosophically puzzling phenomena of self-deception and weakness of the will, in which mental attributions that seem contradictory (e.g., if they are regarded as picking out some underlying state) can instead be seen to issue from different (and nonoverlapping) criteria.

As I suggested earlier, this short study on the different understandings of what it is to act for (such-and-such) reasons needs to be considered alongside Davidson's arguments that reason-explanation is a species of causal explanation. I have argued that this thesis should be resisted, because even if it is accepted that an agent's grasp of reasons may play a causal role issuing in action, this phenomenon is rare; indeed, this understanding is parasitic upon, and cannot accommodate, the two, more central understandings of what it is to act for such and such reasons.

30. Crispin Wright, *Rails to Infinity—Essays on Themes from Wittgenstein's* Philosophical Investigations (London and Cambridge, MA: Harvard University Press, 2001).

Acknowledgments

I have benefitted over the years on the work included here from the observations of many people, including Bruno Ambroise, Daniel Andler, Robert Black, Tyler Burge, Nancy Cartwright, Jim Conant, David Corfield, Tim Crane, Cora Diamond, Frank Döring, Josie D'Oro, Jean-Pierre Dupuy, Hartry Field, John Flower, Philippa Foot, Allan Gibbard, Simon Glendinning, Laurence Goldstein, James Hampton, Edward Harcourt, Jim Hopkins, Pierre Jacob, Jaegwon Kim, Simon Kirchin, Sandra Laugier, Barry Loewer, Cynthia MacDonald, Greg McCulloch, John McDowell, Dick Moran, Richard Norman, Ruwen Ogien, Anthony O'Hear, Christopher Peacocke, David Pears, Helene Buerger Peck, Philip Percival, Georges Rey, Sebastian Roedel, Mark Rowlands, Sean Sayers, Tony Skillen, Murray Smith, Dan Sperber, Helen Steward, Robin Taylor, Charles Travis, Bas van Fraassen, David Velleman, Bernhard Weiss, Ken Westphal, David Wiggins, Jon Williamson, and Crispin Wright. My apologies to those I have neglected to mention.

Since research and teaching are part of the same enterprise, I thank my students for making my job sometimes rewarding and often fun: and to those colleagues, especially Richard Norman, who—respecting my conviction that the teaching environment is a creative one—have encouraged, facilitated, or merely defended my need to keep exploring.

My love and thanks to my mother, Helene Buerger Peck, for making a genuine effort to engage with what I write; to the memory of Mollie for seventeen and one-half years of companionship across three different countries and uncountable homes; to my husband, John Flower, for a lot of things, but especially for demonstrating what it is like to love one's work; and to our "boys," Solutré and Vinzelles, who, with their wonderful and amusing company, make up our family of four.

My respect and affection to the memory of Philippa Foot who taught in the spirit of Wittgenstein without mentioning his name, and to the memory of David Pears, my friend and mentor, who got me interested in philosophy of mind through the topic of irrationality.

This book is dedicated to Crispin Wright for 25 years of inspiration, loyalty, and support as my philosophical interests moved on from there.

Provenance of Essays

Chapter One, "De-Individualizing Norms of Rationality," published in *Philosophical Studies* 79, 3 (1995), 237–258, is a descendant of "Davidson's Characterization of Akrasia," which was presented to the *Groupe de la Recherche sur la Cognition, CREA,* Paris, in 1990, to the Humanities Faculty at Harvey Mudd College in 1991, and published in *Rapports et Documents du CREA,* No. 9009B, in 1990. It is reproduced with permission from Springer.

Chapter Two, "Normativity and Thought," is a response to David Papineau's "Normativity and Judgment" and was delivered at the 1999 Joint Session of the Mind and Aristotelian Society in Nottingham; it appeared in *Proceedings of the Aristotelian Society, Supplementary Volume,* 73 (1999), 45–61. Reprinted by courtesy of the Editor of the Aristotelian Society, © 1999.

Chapter Three, "Playing the Rule-Following Game," was delivered at the Philosophy Departments of Trinity College Dublin and the Universities of Bristol and Hertfordshire in 1995/6, and presented at the European Society for Philosophy and Psychology in Barcelona in 1996. It was published in *Philosophy* 75, 292 (2000): 203–224. Copyright: Cambridge University Press. It is reproduced with permission.

Chapter Four, "Real Rules," commissioned by Jesper Kallestrup and Duncan Pritchard to celebrate the work of Crispin Wright by his students, appeared online as a prepublication in 2008 and in print in *Synthese* 171, 3 (2009): 499–507. It is reproduced with permission from Springer.

Chapter Five, "Why Reasons May Not Be Causes," which appeared in *Mind & Language* 10, 1/2 (1995), 103–126, was conceived as "Reasons, Actions, and Causes," introduced to the Philosophy Department of the University of Nottingham in 1992, then developed as "Reason-Explanation and Mental Causation" and delivered to the Philosophy Departments of the London School of Economics and Political Science and the University of Manchester in 1993. It is reproduced with permission from John Wiley and Sons.

Chapter Six, "Reason-Explanation and the Contents of the Mind," which was published in *Ratio* 18, 3 (2005), 338–351, was presented at the *Institut des Sciences Cognitives* in Lyon and the Philosophy Departments of the Universities of Durham and Exeter in 2003. It is reproduced with permission from John Wiley and Sons.

Chapter Seven, "Reasons as Non-Causal, Context-Placing Explanations," was read to philosophers at the University of Keele in 2007 and published in Constantine Sandis, ed., *New Essays on the Explanation of Action,* Palgrave MacMillian (2009), 94–111. It is reproduced with permission from Palgrave MacMillan.

Chapter Eight, "Pain, Polio, and Pride: Some Reflections on 'Becausal' Explanations," is published here for the first time. An earlier version was presented in 2008 at an Eidos Centre for Metaphysics workshop at the *Université de Genève* and at the Centre for Reasoning workshop on Causation at the University of Kent.

Chapter Nine, "How to Resist Mental Representations," is a critical notice of Tim Crane's *The Mechanical Mind* and appears with a reply by Crane in *International Journal of Philosophical Studies* 6, 2 (1998): 264–278. It is reproduced with permission from Taylor & Francis Group: http://www.tandfonline.com.

Chapter Ten, "On the Conceptual, Psychological, and Moral Status of Zombies, Swamp-Beings, and Other 'Behaviourally Indistinguishable' Creatures," was published in *Philosophy and Phenomenological Research* 69, 1 (2004): 173–186. It is reproduced with permission from John Wiley and Sons. This paper was presented at the University College Cork, Ireland, in 2001 and at a conference on Consciousness and Qualia at the *École Normale Supérieure* in Lyon in 2003. For all those cricketers out there who have asked: Zombie John refuses to name the bowler. . . .

Chapter Eleven, "Conceptual Analysis, Theory Construction, and Philosophical Elucidation in the Philosophy of Mind," is published here for the first time. An early and much shorter version was presented to the European Society of Philosophy and Psychology in Lyon in 2002 and (by Louise Roeska-Hardy, in my absence) to the Wittgenstein Society in Kirchberg in 2002, and subsequently published as "Conceptual Analysis, Theory Construction, and Conceptual Elucidation," in the *Ludwig Wittgenstein Society Proceedings,* Kirchberg (2002): 251–253. Later versions were presented by the author (in English) at the Wittgenstein workshop, Department of Philosophy, University of Chicago and (in French) at the *Université de Paris I, Panthéon-Sorbonne* in 2008.

Chapter Twelve, "Ryle's Regress and the Philosophy of Cognitive Science," appeared in Sandra Laugier and Christophe Al-Saleh, eds. in *J.L. Austin et la Philosophie du Language Ordinaire,* Hildesheim, Olms, 447–467. It is reproduced with permission from Georg Olms Verlag. A version was presented at the Centre for Reasoning at the University of Kent in 2008.

Chapter Thirteen, "Some Constructivist Thoughts About Self-Knowledge," originally appeared as "A Constructivist Picture of Self-Knowledge" in *Philosophy* 71, 277 (1996): 405–422. It began life as "Self-Interpretation," published in *Rapports et Documents du CREA,* No. 9022B, 1990. Early versions were presented to the Universities of Georgetown, Tulane, and Bowling Green: later ones to the Universities of Keele and Nottingham. Copyright: Cambridge University Press. It is reproduced with permission.

Chapter Fourteen, "Self-Knowledge, Normativity, and Construction," was commissioned by Anthony O'Hear for the Royal Institute of Philosophy Lectures Series on Logic, Thought, and Language, which took place in London in 2000. The subsequent publication appeared in *Logic, Thought, and Language,* Royal Institute of Philosophy Supplement 51, Cambridge University Press (2002) 37–55. Copyright Cambridge University Press. It is reproduced with permission.

Chapter Fifteen, "Speaking One's Mind," is a longer version of a review of Dorit Bar-On's *Speaking My Mind: Expression and Self-Knowledge,* which appeared in *Mind* 116, 463 (2007): 727–732. It is reproduced with permission from Oxford University Press.

Chapter Sixteen, "Conceptual Amorphousness, Reasons, and Causes," is published here for the first time. It was initially presented in 2006 as "Reasons and Causes" at the University of Picardie at an international colloquium entitled *Connaissance, Esprit, Valeur* in honor of JohnMcDowell. A shorter version, with a response by McDowell, is to appear as "Amorphie conceptuelle, raisons et causes" in Le Goff and Al-Saleh, eds., *Lire l'esprit et le monde*, Vrin, Paris.

I am grateful to the publishers of these articles for granting me permission to reproduce them here.

Index